Pushed to the Edge

A History of the Naze

And Walton Hall

Ben Eagle

Published by CompletelyNovel.com

ISBN – 978-1-84914-796-5

Printed in the United Kingdom

Book cover designed by Sam Eagle

For Daphne, David and Tigger

And in loving memory of J.W Eagle (1922-2005)

Contents.

Foreword

For more than a quarter of a century, as a representative of the statutory nature conservation agencies (the Nature Conservancy Council, English Nature and now Natural England) I have worked ceaselessly to safeguard the special features of the Naze, Hamford Water and indeed the entire Essex coast.

But that which interests me professionally – the rocks and fossils of the cliffs and the present-day insects, plants and birds of the Naze – are separated by at least a couple of million years. That 'black hole' in my knowledge has now been comprehensively filled by this account, dealing with the human history of the Naze and its environs, our influence on this unique place in historic times, right up to the present, with speculation about our future options and actions.

Fully researched, and written with all the passion of someone intimately associated with the place, Ben Eagle has done the Naze a great service in producing this book. The very act of capturing his sense of place and putting pen to paper keeps it and its particular qualities firmly in the public eye.

I have been privileged to see the genesis of this project from a very early stage. The evolution of Ben's writing style is testament to the qualities of our educational system: from the passionately-held but unstructured views of an amateur, to the rounded analytical approach of the scholar. However it is still highly readable (even to me as a mere scientist), and thus epitomises the interdisciplinary approach he rightly advocates as necessary for understanding now and the future in the context of the recent, historical and much more distant past.

While I may not agree with all his conclusions and solutions for a sustainable future for the Naze, in reality we should not be expected to see things in the same way. Those are the different perceptions of an officer of a government agency – albeit one who woke up each morning for twenty years to see that reassuring reference point, the Naze Tower, from his bedroom window – and someone who has known and loved the place all his life. Different views from different perspectives generate healthy discussion, and if the right decisions are to be made in the future, constructive debate is not just desirable, it is essential. This book deserves to be read by anyone who has the slightest interest in this gem of the Essex coast, and especially by those whose views will shape our collective approach to its future.

Read on for a window in to an Essex few of us knew existed, and still fewer cared about: this book argues persuasively that the future can be truly understood only through the lens of the past.

Dr Chris Gibson

Wivenhoe

CHAPTER 1

Introduction

The Naze peninsula in north east Essex, both the subject and setting of this book, is a place that I have known well throughout my life. I grew up a few miles away from it and during my childhood it became one of my many playgrounds of wide open spaces and, in hindsight, a key learning landscape. Gradually I developed a keen interest in the landscape itself and wanted to find out about the stories hidden within its past. It is the subsequent investigation led by this initial curiosity that resulted in this book. However, although my inspiration may come from personal experience the agenda of the book is driven by my determination to write from a 'detached yet connected' point of view. 'Detached', because objectivity of source material should be king when writing history but 'connected' in displaying a passion for the place that I hope will be evident throughout. During the course of writing the book my perspectives have altered as my experience has broadened. I have come to know the many different aspects of the Naze landscape and to begin to understand the subtle influences it has had on local culture and social history. From the vast expanse of marshes on the western side of the peninsula to the sandy beaches to the north and the famous cliffs on the eastern side, all have their peculiar identity and all have shaped the history of the place in their own way.

My historical training has been critical in developing the ideas presented through the book. I studied history at Bristol University under the supervision of historians including Peter

1

Coates, Andy Flack, Ian Wei, Ronald Hutton and Brendan Smith. This broadened my perspectives and opened my eyes to grand narratives and social and cultural theories that shape places and their pasts. When I began putting the book together in the years before undertaking my degree I had not been exposed to such ways of thinking and therefore the manuscript you are reading today is very different to the book I originally put together several years ago. The previous version remains important to me as a record of my pre-graduate writing style but the current version, I hope, provides a little more in terms of reflection and historical analysis. For various reasons, including developing my interests outside of my direct disciplinary field, taking part in practical environmental conservation, becoming increasingly interested in the natural sciences as well as being inspired by members of my own department and reading around the subject, I came to specialise my studies in environmental and animal history.

Environmental history is an interdisciplinary subject, marrying the social and natural sciences with the arts and humanities, among other disciplines. It is this keen interdisciplinary interest that has, in writing this book, drawn me towards an approach that takes in to account both the historical and environmental peculiarities that the Naze presents and how these interact. I find the Naze fascinating because of its environmental profile, its volatility as a landmass and the aesthetic of its surrounding but also the cultural and social aspects of its history and present situation. Whilst there is analytical material available documenting much of the prehistoric, geological and natural history of the peninsula there has been little work on the human history of the Naze, with the exception of a number of short theses by the most prolific chronicler of Walton's history, Peter Boyden. There has been some 'questionable' material that I have come across during my research (unreferenced works abound) but I have tried to include as much existing material that seems reliable and relevant as possible so that previous study of Walton and the Naze can be scrutinised.

I should say at this point that my personal links with the Naze go further than my childhood experiences. My family has farmed at Walton Hall, the only farm on the peninsula, since the 1880s when my great great grandfather moved to Walton from Wix Abbey, about 7 miles as the crow flies from the town. This particular familial relationship with the place has been useful in researching for this book not only because I have been able to easily access a large number and variety of sources relating to Walton Hall farm (which makes up more than 2/3 of the Naze's total area) located in the farm office 'archive' (in reality a chest full to the brim with old accounts, maps, plans, letters and photographs), but I have also been able to freely access the land itself ever since I was a small child. More recently I have worked for Essex Wildlife Trust at the Naze, a period in which I have been able to explore it in greater detail and speak to many people about it, learning from their own experiences and knowledge. The book itself is the result of an amalgamation of primary research, secondary reading, engagement with previous local histories, local people and historians as well as undertaking oral history interviews and conversations I have had along the way.

A key theme throughout the book is coastal erosion and issues surrounding how coastal processes affect landscapes and settlements. The Naze, made up of soft geological features, is eroding at a rate of 1-3m per year and its situation is widely acknowledged as an example of the pressures that coastal erosion places on local communities. In 2011 the 'Crag Walk' rip-rap coastal defence project was completed with the intention to halt, or at least reduce, erosion to the south of the Naze public open space. The Naze is valued by geologists and naturalists, interested in the peculiar rock formations, fossils and habitats, birdlife and plantlife, to dog walkers and day trippers. It is a key economic asset for the town of Walton. As a symbol for the town and a key tourist attraction, the town has much to lose if the Naze were to lose its accessibility and its economic viability. This cultural connection, whilst not entirely overlooked in previous histories of the town, has not previously been discussed at length. It will be made clear in this book that culture as well as natural history drives the agenda for the reasons behind the Naze's depiction as an 'important asset'. Over 100,000 people visit the Naze each year, many of whom are students looking to explore these issues. I hope that this book can add to the tools available to these students to help understand the Naze in the context of its human history as well as its present social struggles and its geological history.

The 'old' town of Walton lies to the east of the concrete sea walls that today protect the settlement from regular flooding. The Naze lies to the north of the central part of Walton, although over the past hundred years the headland past the 'Walton gap' has become increasingly developed and has lost its 'isolationist' and more rural identity. The Naze has multiple identities which will be explored in the book. It is known by many as a fundamentally important example of Pleistocene and Eocene geological formation but it is also seen as a place 'under threat' from the sea, in need of 'protection'. Some suggest that this 'fragile' peninsula is being, to use the title of this work, 'pushed to the edge'. Perhaps due to my historical training I am wary of any term that suggests that an environmental change is necessarily a good or a bad thing or that environmental processes are driven by intention. Clearly the sea is not consciously eroding the Naze, even if the effects may be perceived as damaging for local people. A principal difficulty faced by environmental historians (and historians more generally) comes in the use of language and ensuring that the words used do not imply particular meanings that may affect the way that histories are perceived. Semiotics is a key problem when writing about erosion as there is a risk of viewing the sea as a 'threat' that is 'claiming' the Naze for 'itself' whereas, although the rate may be influenced anthropogenically, it is simply a 'natural' process.

The book has been titled 'pushed to the edge' in reference to the fear by some that the landscape will soon be 'lost' and that this is fundamentally a bad thing. I will suggest that landscapes should be seen in historical and geographical context and future changes should not necessarily be feared but embraced and adapted to. Many people fear change itself which is rarely necessarily either a good or a bad thing. It is much better to take an attitude of developing mitigation and adaptation and to accept that future generations will adapt to the situation with which they are presented. Set within this must be a degree of conservation in terms of the maintenance of the historic landscape. It is landscape that provides a relative constant in time and can bind generations. Oliver Rackham described landscape as:

> a historic library full of 50,000 books. Many [landscapes] were written in remote antiquity in languages which have only lately been deciphered; some of the languages are still unknown. Every year fifty volumes are unavoidably eaten by bookworms. Every year a thousand volumes are taken at random by people who cannot read them, and sold for the value of the parchment. A thousand more are restored by amateur bookbinders who discard the ancient bindings, trim off the margins, and throw away leaves that they consider damaged or indecent...The library trustees, reproached with neglecting their heritage, reply that Conservation doesn't mean Preservation,

that they wrote the books in the first place, and that none of them are older than the eighteenth century.[1]

Sadly most of us fall to the mind-set of shifting baseline syndrome and see the world only through our own generational perspective, often ignoring perspectives that the past can present us with. This is one of the great applied benefits of reading and writing history. History can encourage us to look beyond our 'baseline' and understand landscapes and the environment more broadly in the context of change, adaptation and the lives of past people, animals and plants.

During one of the interviews I carried out as part of the research process, the interviewee effectively illustrated the reality of the present situation by saying that although land is being eroded at the Naze it is building up further south, on the north Kent coast. This particular person, unlike many, took a broader perspective and saw the need to accept change and to adapt to it rather than prolong the inevitable and avoid planning for a sustainable future. Landscapes must be seen in the context of wider systems and we must learn to adapt to those systems. This does not necessarily mean that landscapes such as the Naze should not be 'protected' but such protection should be envisaged within a context of adaptation in the longer term. It is important to have a back-up in landscape and urban planning. Eyes must face forward not backward, but it is also important now and then to take a glance in to the past to understand change on a longer term basis.

Driven by an association of past connections landscapes have many different 'value' associations: historically, socially, culturally, ecologically and economically. These values are associated with meanings and affect the identity of a place. Once these perceived values are removed the identity of the place suffers. The Naze must be seen within the context of these different but associated values and in the context of its landscape - locally, regionally and globally. In the most local vicinity the headland lies at the mouth of Hamford Water, or the Walton Backwaters as it is known colloquially. Hamford Water is an embayment, known as such because it does not have any major freshwater inputs, of over two thousand hectares in area, and contains a great extent of salt marsh. The Backwaters consists of mud flats, saltings and various islands. It was designated a site of special scientific interest (SSSI) in April 1961 due to its peculiar environmental condition and has been immortalised as the setting for one of Arthur Ransome's famous novels, 'Secret Water', first published in 1939. For Ransome's *Secret Water*, Ursula Bloom's *The House that Died Alone* and Paul Gallico's *Snow Goose*, the Naze was an important feature in their storytelling as it forms the eastern boundary of their Backwater fictional works. The Backwaters, like the North Sea to the east, are fundamental to the Naze's spatial identity.

Regionally and globally the Naze has historic and natural historic value as part of the wider geological landscape. It also has social, cultural and economic value due to the vast associations with other places that will be explored throughout the book. These particular associations reach out to the heart of London in the case of its historic associations with St Paul's Cathedral, to Scandinavia, Normandy, the north-west of England and to the wider continent, particularly relevant in its relationship with its neighbour across the water, the maritime town and port of Harwich. The Naze continues to attract visitors from around the world drawn by the seaside culture of Walton-on-the-Naze, a phenomenon encouraged and developed in the Victorian and Edwardian periods and that is just about surviving today, albeit having declined substantially from its heyday a century ago, as well as the Naze Tower, the Naze cliffs and the resident wildlife. Globally, the Naze will be shaped by the phenomena of climate change

7

and sea level rise, changing conditions to which we must learn to adapt to, not to fight against.

We live in an age of the STEM subjects (Science, Technology, Engineering, Mathematics), a time when the value of the arts and humanities is increasingly under pressure in a society obsessed with utility and practical application. However, history can be useful in providing an analysis of past events in order to aid future decision making and give present decision makers information about previous actions. Environmental history in particular will be crucial in the longer term in understanding the climatic and broader environmental change we have experienced and will experience in the coming years. However, to be able to adapt to future situations in the way that I am calling for, it is vital that the value of all disciplines is utilised. The most exciting research will be the studies that work on the boundaries of their own discipline with other disciplines, collaborative work that takes in to account both scientific and socio-cultural perspectives. The Naze provides a case study whereby scientific study could effectively sit side by side and interact with arts and humanities perspectives for planning a sustainable and meaningful future for the people who live there and the people who visit. The role of the humanities is partly to analyse value structures embedded within narratives and places. Popular historical works (perhaps unlike academic historical works), such as this, have an important role in generating and enthusing public consciousness surrounding issues and have a duty to be accessible. This historical assessment of the Naze therefore aims, at least in part, to raise public awareness of the historical significance of the peninsula and, in turn, the cultural, social, economic and natural historical implications. It follows a grand tradition of local history writing but it is embellished with case study analysis, taking a chronological structure at first but then shifting thematically and geographically to study the Naze Tower, Walton Hall buildings and the relationship between landscape, sea and people over time.

Chapters two, three and four take a chronological approach and place the Naze's history in context, giving the reader a broad overview in understanding the landscape and the place's connections with other places and people over time. The second chapter briefly explores the geological profile of the Naze as well as explaining developments in the wider region through the Bronze Age and Iron Age as well as the Roman and Anglo-Saxon histories of the region. The role of St Paul's Cathedral has been significant in shaping the identity of the peninsula and its status of special importance within the 'Soke' and so its role, as well as the early development of the Church, are discussed in the third chapter. The chronological starting point for the fourth chapter is 1544 and it centres on the theme of landownership and control, demonstrating how this new narrative of 'detached' aristocratic landownership followed by owner-occupier landownership influenced the Naze's identity. The fifth chapter is centred on the subject of Walton Hall house and curia[1] and chapter six on the Naze Tower. Chapter seven explores the subject of the Naze's relationship with the sea and then the conclusion engages with the idea of 'saving' the Naze for posterity and the cultural and historical roots of a landscape perceived by some as being 'pushed to the edge.' In the final section, set out as an appendix, a series of oral history interviews are notated, all of which were conducted during the course of researching for the book. They provide a uniquely personal method of explaining the issues and stories involved at the Naze, exemplifying the strong attraction this small peninsula has for people.

---

[1] curia – the principal buildings of the manorial complex, surrounding the manor house

The book aims to inspire interested locals and visitors to understand the Naze in the context of its history as well as attempting to answer why, from the 1970s, the formal movement to 'save the Naze' came about. A key finding of my research has been that, generally, the issues that decision makers have faced over time at the Naze have had similarities between them. The majority of decisions have had the sea at their forefront. A particular continuity has been a determination to keep the Naze predominately a place of human industry and agriculture, trying to avoid what is perhaps the inevitable: it 'falling' to the sea. Agriculture remains the majority land use for the peninsula (arable and grazing), although foremost in most people's minds is the current management of the eastern edge of the landscape – for leisure and conservation purposes on the 'public open space' and nature reserve. When Essex County Council bought the former golf club site in the 1960s the Naze became coupled with a 'grander' purpose – to be a 'public open space' for local people and visitors to enjoy. When the sea erodes this landscape it is also seen to threaten the community that 'owns' it and so it gains its own distinct identity.

The Naze is currently predominately owned by the J.W. & F.D. Eagle farming partnership with other landowners including Anglian Water, Essex Wildlife Trust and Tendring District Council. On the eastern edge of the Naze, the historic Naze Tower, now privately owned, was, until the building of the 'Crag Walk' within a decade of falling over the cliff edge. The Crag Walk coastal defence system was completed in 2011 and is an 'educational public viewing platform' set upon a structure of boulders that dissipate the wave energy. The project was spearheaded by a number of organisations with a keen interest in the Naze including Essex Wildlife Trust, the Naze Protection Society, the Naze Tower, Essex County Council, Tendring District Council and Natural England. The Crag

Walk project is seen as the answer to slowing the erosion rate at the Naze as well as saving the Naze Tower from falling over the edge. The defences will not however stop the erosion in its tracks and due to drainage issues the cliff will continue to retreat, albeit at a predicted rate of twenty metres over the next seventy to one hundred years, compared to the average retreat of one to three metres previously experienced and still experienced elsewhere on the peninsula. People and land managers will have to adapt to future situations as they arise.

My generation will see dramatic changes to the natural environment, both globally but also locally. The world faces several key challenges over the next few generations, including human population growth, the issue of how to feed that growing population, biodiversity loss, how best to move towards a fossil fuel free economy, and connected to all, climate change. The Intergovernmental Panel on Climate Change (IPCC) issued their fifth Assessment Report in 2014, predicting that by the end of the 21st century global temperatures will have warmed by at least 1.5 °C. Sea levels are also predicted to rise substantially, reflecting past recorded figures. Over the period 1901 to 2010, global mean sea level rose by nineteen centimetres and the rate of sea-level rise since the middle of the 19th century has been larger than the mean rate during the previous two millennia. Looking ahead to 2100, model projections for global sea level predict a rise of anything between twenty eight and ninety eight centimetres, with regional fluctuation. For an environment such as the Naze and for the surrounding Hamford Water National Nature Reserve more widely, these predictions would mean a complete alteration in terms of landscape, ecology and economy. Coastal change takes place as a result of a combination of different factors, including geological factors (nature of rock type, structures within rocks, position in the water table, changes in the earth's crust), climatic processes (waves, rainfall, tides, sea

11

level, storm intensity) and human factors (infrastructure such as harbours, aggregate dredging, coastal defences). Coastlines are dynamic entities that can either retreat or accrete and in a modern age where the first response is usually to solve a problem technocratically we must learn to adapt to the situations we will find ourselves moving towards.

Weather patterns are also predicted to become more unsettled. When my great, great grandfather came to Walton Hall in the 1870s there were infrequent storm surges such as that in 1897 that sometimes caused significant damage to the sea defences. The upkeep of sea walls has often been a contentious issue in the terms of leases, particularly in the case of landlord Mrs Welch and tenant John Hicks in the 1820s. Since then, there have been a number of significant coastal flooding disasters along the east coast of England, most notably in 1953 when much of coastal Essex, including large stretches of the Tendring coastline, was overwhelmed, resulting in the devastation of farmland and the loss of many lives, especially in Jaywick and Canvey Island. Storm surges are predicted to become more frequent in the future. The sea walls built and secured following the 1953 flood will not be able to cope with these problems and are also under threat of being undercut by erosion to the north east corner wall of the Naze.

The situation of the Naze is interesting as a case study representing one of many dozens of similar cases up and down the British coast. It is often used by schools as an example for the effects that coastal processes, coastal erosion and wave energy have on land and settlement and it is cited within the geography GCSE syllabus as a potential case study for coursework purposes. Coastal erosion threatens the built environment including domestic, commercial and industrial buildings and infrastructure. Many of our prominent towns and cities, as well as industrial port complexes are situated along coastlines and estuaries. The concentration of resources in coastal areas means that it is imperative that we find short and long term solutions for each area of the coast if these settlements are to be used in the future. Shoreline Management Plans (SMPs), although not binding, exhibit a vision of this process. SMPs broadly recommend where erosion can be checked, where the line can and should be held and where retreat could be an option. However, there will always be debate as to what is worth 'saving' and what should be left alone. 'Value' is an entirely subjective entity.

I want to briefly illustrate two case studies of areas facing similar situations to the Naze. One is the Holderness coastline in Yorkshire and the other the Isle of Wight. The Holderness coastline is predominately made up of cliffs of twenty to thirty metres high, consisting of soft boulder clay. Where the North Sea meets the Humber estuary a spit has formed, known as Spurn Head. Erosion rates are on average between one and two metres per year. Most of the material that comes from the cliffs is washed out to sea, with the rest moving southwards through the process of longshore drift. Critically, Easington Gas Station is situated on top of the cliff. It is crucial that its position should not be compromised for various safety reasons. This is similar to the Naze which has a low lying water treatment works within the landscape. A number of defence mechanisms have been tried in the past at Holderness including the construction of groynes along the beach, the construction of a sea wall to protect Easington and the construction of off shore breakwaters in the form of tyres and concrete blocks, forcing waves to break offshore. Crucially, and this is relevant to the Naze and the north Essex coast, only the most valuable areas of land are protected due to the extensive costs of defences.

This is also true on the Isle of Wight. The 110km of coastline on the island are extremely varied and the erosion rate is dynamic. The cliff line is nearly continuous but weathering and erosion rates vary throughout, especially between the particularly vulnerable south west coast and the relatively stable north coast. Again, decisions upon where to hold the line and where to retreat it or whether not to take action at all depend very much on perceptions of 'value' and, crucially, where it makes economic sense to defend the line. Hard defences have been built at Cowes, Ryde, Ventnor, Sandown Bay and in the North West, near Bembridge Airport. However, the construction of hard defences in some places will affect the natural shoreline dynamics elsewhere which may have even greater implications. Farmland and areas where there is little built development are perceived as being less 'valuable'.

Both of the above examples raise important questions relevant to the situation at the Naze. Why 'protect' some areas more than others? How should each area be 'protected'? How can 'protection' be justified? Ultimately it is policy makers who must decide not only which stretches of the UK coast should be most heavily 'defended', but also undertake the most appropriate form of defence for a particular locality. This is where the debate between 'hard' and 'soft' engineering methods comes to the fore. 'Hard' structures such as groynes and sea walls may prolong the inevitable or actually increase the rate of erosion. 'Soft' engineering processes broadly work 'with nature', such as by encouraging new salt marsh creation as a means of dissipating wave energy. However, such methods mean a fundamental change in land use, often difficult for landowners, land managers, decision makers and the wider public to accept. The Naze is valued in different ways by different people and some argue against the construction of hard defences in front of the Naze cliffs for several reasons including the prevention of any future fossils from emerging, prevention of material moving south to reduce erosion further down the coast and reducing the value of the habitat for sand martins. The Crag Walk, they say, is as far as it should go. Others believe Crag Walk should be extended and defences strengthened throughout the peninsula, holding the line for as long as possible. This is a risky strategy and does not, in my opinion, allow for a sustainable land use strategy for the future. Different agendas must be taken in to account but ultimately it is up to funders and governments to decide whether to take a utilitarian approach or to favour the point of view of a particular minority interest group. The main issue is that people will have to work together, moving in the same direction in forging a future vision for the Naze.

When looking at coastal management, it is imperative that sections of the coast should not be managed in isolation. When coming to conclusions about defence solutions, answers must relate to the bigger picture. Changes in the past, present recorded observations, future models and analysis of situations elsewhere must all be taken in to account. Sustainability should lead the agenda and an integrated coastal approach taken, looking towards mitigation and adaptation. I hope that this book will help decision makers in their discussions about the future of the Naze. I hope that stakeholders will work towards an overall considered, sustainable and viable plan for the Naze headland as a whole, rather than focusing on piecemeal aspects and the shorter term. The Naze should be seen in the context of other places but its own value should also be highlighted. 'Saving' the Naze may be impossible in the longer term but a plan to adapt to a range of future potential situations will provide for a more positive and sustainable vision.

CHAPTER 2

Heritage and Prehistory

The Naze has been both physically and culturally influenced by the North Sea, an entity that must be engaged with throughout the telling of this history of the Naze. The famous cliffs themselves have been defined over time in relation to the sea. I am by choice a landlubber, someone whose first instinctive thought is perhaps to 'protect' the land as the 'higher partner' in a hierarchy of landscape. However, as I have come to better understand the Naze (and better understand the nature of environmental history) I have come to appreciate marine environments and my previously 'defeatist' mentality has changed. Instead of seeing a loss of landscape I see a transformation, from dry to wet, from terrestrial to marine. Nonetheless, today the Naze is visibly and rapidly changing. This is not a new phenomenon, as will be explained through the course of the book, but it concerns many people and organisations who have campaigned to 'save' the Naze for years, something that will be analysed later. This first chapter puts the relationship between the Naze landscape and the surrounding seascape into context, explaining how settlement began to develop in the wider region, looking at migration, food production and other human influences on the environment. The sea itself will be addressed directly in a later chapter. Although the book is predominately concerned with the human history of the Naze, principally over the past 15,000 years, passing from the Pleistocene to the Holocene and the Anthropocene, natural history revolves intimately within this context. A brief explanation of the geology of the headland is perhaps therefore necessary to explain the deep historical and geological context for the erosion experienced today.

The landmass is made up of various different layers of sedimentary materials, specifically London Clay, Red Crag and a layer of sand and gravel deposits towards the surface. Rain permeating through the upper layers of topsoil, sand and gravel, today, acts as a lubricant and thus material at the top slides over the underlying clay base, resulting in 'slumping', the major reason for the rapidity of cliff erosion on the site. This 'slumping' happens in conjunction with the erosion at the base of the cliff, the result of swash and backwash, destabilising it.

The known geological story of Essex begins with rocks that are between 440 and 360 million years old, dating from the Silurian and Devonian periods. These rocks are over 300 metres beneath the surface and consist of shales, mudstones and sandstones. Lying on top of these is a marly clay known as the Gault, dating from around 100 million years ago. It should be noted therefore that there is a gap in the geological record of Essex of about two hundred and fifty million years. Nonetheless, a hundred million years ago sea levels were rising and this set the conditions for the next rock to form: chalk, which can be seen on the surface at areas in the south and west of the county such as in Thurrock. Chalk forms the base of the 'London Basin' and was laid down as a limy mud on the bed of a tropical sea whilst dinosaurs roamed the Earth. This chalk sea covered Essex during the Cretaceous period, between eighty and a hundred million years ago. This sea teemed with marine life from molluscs to corals, fish and marine reptiles up to ten metres in length. Microscopic marine algae accumulated on the sea bed.

The London clay beds, visible at the Naze today and estimated at being a hundred and seventy metres thick, began to form in the early Eocene period, circa fifty million years ago, under a tropical sea about a hundred metres deep. The clay originated as a layering and compaction of silt and mud deposits delivered from rivers flowing from inland. Northern Europe had a climate comparable to modern day Malaysia and crocodiles, sharks and giant turtles swam in the local waters. Looking for sharks' teeth is a popular activity at the Naze today, as it has been for many decades. Britain was a land of tropical mangrove and palm forests. The rivers also carried wood, fruit, seeds and bones of mammals, all of which have been preserved and fossilised within the London clay formation.

The Red Crag layer, a profile of sand stained red by iron oxide, is packed with hundreds of shells. Indeed, nearly three hundred species of shells have been found at the Naze, famously including *Neptunia contraria*, the left handed whelk, spiralling the opposite way to modern gastropods.[2] The stratum, which occurs across much of north Essex, albeit only well exposed at Walton, was laid down as dunes on the sea bed from the Pliocene in to the Pleistocene (2.5 to 1.5 million years ago), before the last Ice Age. At the time the Naze lay beneath a shallow sea of fifteen to twenty five metres and the climate remained warm. The Red Crag layer is of the Waltonian stage and represents the oldest deposits of the Pleistocene in Britain. Critically, the layers within the Crag alternate with colder and warmer water fossils, indicating climatic fluctuation during this period, ultimately resulting in the last Ice Age. The Red Crag at the Naze is the earliest accessible dateable evidence of the onset of the last Ice Age in North Western Europe, and is thus used as a baseline date for all

---

[2] A gastropod is any mollusc characterised as having a single, usually coiled shell or no shell at all, a ventralmuscular foot, and eyes and tentacles located on a distinct head. Includes snails, slugs, limpets and cowries.

subsequent geoclimatic events. Today, sand martins can be seen nesting within the cliff or feasting on insects on top of the headland. Rabbits also burrow within the cliff. The sand and gravel layer on top of the red crag is the result of fluvial transportation. Six hundred thousand years ago, during the early ice age, the river Thames flowed through north Essex, Suffolk and Norfolk, making its way across the Doggerland (what is now the southern North Sea) as a tributary of the Rhine. The River Medway also flowed north across East Essex to join the Thames near Clacton, leaving behind a ribbon of distinctive gravel deposits. The Pleistocene epoch consisted of a number of glacial and interglacial periods before the temperate Flandrian stage of the Holocene came about and the River Thames shifted to its present route.

As mentioned above geologists recognise that Essex, along with the rest of eastern and southern Britain, was once adjoined to the European continental landmass and this vast expanse of land is known as the 'Doggerland'. [2] During the most recent glaciation, a period that ended around eighteen thousand years ago, the sea level was about a hundred and twenty metres (390ft) lower than today. The island of Britain was yet to be formed and the area between the Netherlands and Scotland was a vast expanse of low lying tundra composed of shrubs, sedges, grasses, mosses and lichens. The northern coastline was one of saltmarsh, lagoons, mudflats and beaches, similar to the habitats found at the Naze today, albeit on a vast scale. It is thought to have been a rich place for hunting, fowling and fishing. The Holocene Climate Optimum was a warm period c.7000BC to c.3000BC when an increase in temperature of around four degrees Celsius took place near the North Pole. Although southern Europe experienced relative cooling, north-western Europe experienced significant warming, with up to a six degree rise occurring in Britain. Significant environmental changes took place during this warm period, none more significant than in Britain, which became an island after the Dogger Bank was flooded.

In 2008 Bernhard Weninger et al suggested that around 8000 years ago (c.6200BC) much of the Doggerland was flooded by a tsunami, resulting from a submarine landslide off the Norwegian coast.[3] This is known as the Storegga slide and involved the collapse of 290 km (180miles) length of coastal shelf with a combined volume of 3,500 km^3 of debris. More recently, it has been suggested that the release of the freshwater from the glacial lake Lake Agassiz in North America, between 13,000 and 8200 years ago, was a cause of the flooding of Doggerland. No matter the reasoning it is clear that the identity of the Essex landscape changed significantly during this time. It was the first time that Britain was an island.

There is evidence of human activity in north Essex from at least fifteen thousand years ago. Sculpted flints dating from around 13,000 BC have been found in the Colne Valley and at Shoeburyness.[4] In 1888, an amateur archaeologist, H. Laver, found a polished stone axe and several arrow heads at Stone Point, at the northern most tip of the peninsula.[5] Arrow heads have been found on the beaches leading up to Stone Point since then. It is likely that due to the significant rise in sea levels of about thirty metres or so since the Palaeolithic, many of these materials remain buried under large amounts of sediment. Whilst many people today look negatively on the retreat of the Naze cliffs, principally due to its effects on the landscape and potential land-use implications, fossil hunters could potentially make significant discoveries as the cliff continues to recede.

Samuel Hazzledine Warren was an amateur geologist and archaeologist from Loughton who was particularly interested in the London Clay at Walton and was working in the area of the coast in the early 1900s. One of Warren's most exciting discoveries was in September 1910 when he found an almost complete prehistoric human skeleton (considered to be a male) near Stone Point, with his companion Miller Christy. Dr Arthur Keith, an anatomist at the Museum of the Royal College of Surgeons carried out the preliminary report in to this specific discovery. He wrote to Warren as follows:

The particular value of the prehistoric skeleton you have had the good fortune to discover at Stone Point, Walton-on-the-Naze, lies in the fact, firstly, that it is perhaps the most complete prehistoric skeleton ever discovered, and secondly, that the data relating to its antiquity have been observed and gathered with accuracy. When you first showed me the delicately moulded and rather diminutive skull I gave it as my opinion that it must be that of a woman; but as you pointed out, the pelvis left no doubt as to the sex: it is that of a man. My preliminary mistake emphasises the character of the whole body – it is finely moulded, slender, feminine in type. By actual measurement I estimate the height to be 1592mm (5 feet 2 ¾ inches), and by comparison with skeletons in the Museum there can be little doubt that the estimate is accurate. He was a young man: the wisdom teeth are just coming in to use, and the last epiphyseal lines of the long bones closing, so we may regard him as about 23 or 24. He was a rather slender and dapper young man, but finely formed, straight and well proportioned. His legs are straight and symmetrical, and bear a normal proportion to arms, trunk and head...He is just a little fellow, such as one may see busy in the offices of the city today...Another interesting point is the marked specialisation of the right shoulder and arm. He was undoubtedly right handed. I have never seen such a

21

specialisation of the right shoulder joint. Whatever trade he followed was one which entailed a constant rotation of the right shoulder, a twisting out of the whole arm such as one employs in using a corkscrew or a cobbler stitching. Could he have been a weaver? ...[6]

In Warren's study of Palaeolithic deposits at Walton, written in 1918, he pointed to two significant discoveries, both of large mammals. First is an early record of 1610 from Camden's 'Britannica':

> What hath been found in this place, have here out of the words and credit of Ralphe, the monk of Coggeshall, who wrot 350 years agoe. 'In King Richard's time, on the seashore, at a village called Erdulphnesse (Walton), were found two teeth of a certain giant, of such a huge bignesse, that two hundred such teeth as men have now a daies might be cut out of them. These I saw at Coggeshall.[7]

The second is a record from 1803 denoting a fall of the cliff which exposed the skeleton of a 'Mammoth' measuring 30 feet in length. Two molars of the mammoth were also found and weighed 7 and 12 pounds. It had been deposited about a mile SSW of the Naze, near the old Bath House hotel.[8] It is difficult to map the extent of the area of the Naze over time due to the degradation of soil materials and their dissolution in the sea. The only material evidence we can make use of are the small number of basic tools found on the beaches on the eastern side of the peninsula and at Stone Point to the north, showing the variety of activities that were being carried out at various times and where these were happening.

After the Cold Period of the Younger Dryas (between 10,800 and 9500BC) Essex became engulfed by hundreds of pine and birch trees which colonised the former steppe-tundra. As the climate continued to warm pine forests gave way to broadleaf species, the varieties of trees that we are used to seeing in woodlands today, Oak and Ash especially. This affected mammal and invertebrate species diversity but humans were able to adapt to the situation, exploiting the species available. During the later Neolithic period (circa 3,500BC – circa 2,500 BC) there was a further distinct change in climate which affected social practices. Temperatures were beginning to fall by the mid fifth millennium BC although sea levels continued to rise, albeit more slowly.[9] As the Stone Age passed into the early Bronze Age the Essex landscape changed substantially. Towards the end of the Neolithic (c.3000BC), deciduous woodland covered many parts of Essex with trees running down to the seashore.[10] In the Bronze Age however, and continuing into the Iron Age, forests were cleared, settlements were created and, perhaps most importantly, structured farming developed. People began to concentrate efforts on growing crops and domesticating herds rather than relying on gathering food and hunting, removing the 'pot luck' element to survival as much as possible.[11] The landscape changed on a huge scale as woodland was felled and replaced by grassland for grazing and primitive cereal crops. Grain was collected by the community and stored to last through difficult winters. Animals such as oxen, sheep and dogs were bred, both as a 'resource' for food and clothing, as a labour force and as companions.

Flint exploitation became more widespread. There is large scale evidence of microliths – small flakes of flint fashioned to fit on wood – in the area of Walton which suggests that the site around the Naze was a manufacturing area for tools and weapons, traded widely throughout the Neolithic.[12] Further evidence of the ingenuity of this society comes from the evidence of a small scale boat building industry. Flints were used to hollow out tree trunks to make 'dug out' boats, ten feet long and three to four feet wide, used to dredge oysters and other shellfish for human consumption. In 1936 a boat was found above Lion Point, near Jaywick, dating to around circa 4,000 BC.[13] The boat building industry supported trading networks, conducted along long established routes.[14] Bronze led to the improvement of spear technology with the addition of sockets to spears, into which could be forced a shaft and spearhead.[15] Rising sea levels affected the scale of agricultural activity as there was less land available for grazing, meaning that the major industry in the late Bronze Age was fishing, not agriculture, fish being a vital food resource in an age of food insecurity.[16]

The Iron Age came to Essex circa 550 BC, lasting, historically, until the coming of the Romans in AD 43. The expansion of settlements in the area of north Essex combined with the industrial and agricultural activities at the coast indicate that the Naze area was part of a wider integrated network, although population remained very small. The main preoccupations of most of the population were raising sheep and cattle, as well as agriculture, similar to the Bronze Age.[17] However, the principal crops grown had changed by the late Bronze Age. Emmer wheat (*Triticum dicoccum*) and naked barley (*Hordeum tetrasticum*) had given way to Spelt wheat (*T. spelta*) and hulled barley (*H. hexasticum*), mainly due to the changing climate, which favoured these autumn sown varieties which would ripen earlier than those sown in spring.[18]

For the majority of the Iron Age, north eastern Essex was settled by the Trinovante tribe, who occupied the eastern parts of the modern county during the pre-Roman period. However, they faced a wave of invaders from the Continent, including the La Tène people and, from circa 75BC, the Belgic tribes.[19] The Trinovantes were mainly indigenous Iron Age farmers,

24

although Gallo-Belgic coins that have been found in north Essex and the Welwyn type graves indicate that there was also a Belgic element, possibly a ruling class of warriors or adventurers who had recently arrived from Northern France.[20]

Before the Romans came to Britain there was a relatively sophisticated economic structure in place in Essex. Iron Age farmers had considerably altered the landscape through creating field systems that the Romans later claimed to be one of their so called 'civilising' legacies. From pollen analysis and carbonized seeds found in excavations, it is known that Emmer, Spelt, barley and oats were the most commonly sown Iron Age crops, barley being the most important.[21] Many new trade networks were established during the Iron Age and people were experts in exploiting the raw materials they found locally. Salt extraction was another important industry, salt known as 'the gold of the Iron Age'. Several 'red hills' have been found in the area to the south of Hamford Water, to the west of the Naze, evidence of an ancient salt extraction industry. Tool and pottery making were also expanding industries and developed at an extraordinary pace.

Influence from abroad came through trade, not conquest. The Naze was part of a trading network across several 'empires of influence'. In the first century BC the Belgae people spread across a large amount of southern England, absorbing smaller tribes as they went. They introduced several ideas and technologies, including the wheeled plough, which greatly improved agricultural productivity.[22] The leader of the Catevellauni tribe, of the Belgae people, Cunobelin, was a particularly successful leader and developed his capital Camulodonum (Colchester) to become am economically significant commercial centre. It was Cunobelin who founded Camulodonum, the 'stronghold of Camulus, the war-god', in the first century AD.[23] During his leadership period the

Catevellauni defeated the Trinovantes, inflicted defeats on the Iceni to the north and by the end of his reign had merged several peoples to create a wider Belgic sphere of influence.[24] Cunobelin's reign marks what can be noted as a division of eras – from the rule of indigenous tribal people towards the conquest of Roman Britain.

Whilst Camulodonum remained the most important administrative town in Roman Britain, the area around the Naze was not much more connected, in terms of infrastructure, to the surrounding area than it had been prior to the invasion of AD 43. The major road networks spread from north to south linking Camulodonum to Londinium (London) and east to west linking Camulodonum with Verulamium (St Albans) but links to the coast were difficult, making settlement there very remote. It appears there was some access over land to the coast although none were major routes. Canon J. Allen has suggested that 'one chronicler' noted that a Roman road existed, starting at Beaumont Quay and leading to Colchester through the 'wild countryside' of Tendring.[25] Evidence for the reasons behind the construction of this road comes from industrial activity which needed transportation links to the capital. The major industry at the Naze in the first century AD was the collection of septaria, an argillaceous limestone found in London clay, used as a building material (sometimes known as Roman cement) and used when building the Balkerne Gate in Colchester. M.R. Hill, curator of Colchester castle museum in the 1940s, believed that septaria was taken by barge to what is today Beaumont quay, from where it was transported through Tendring and Hare's Green using an elongation of the Roman road which ran from Elmstead to Colchester.[26] For the majority of the Roman period the economy of the Naze remained agricultural. Spelt and Emmer wheats, along with barley and oats made up the arable crops with sheep and cattle pastures forming the rest of the landscape. Sheep breeds were something akin to the Soay or Shetland, primitive and small, and many hands were needed to shear them.[27]

The Roman army remained in Essex until the 5th Century AD but as central authority in Rome waned Roman Britain became increasingly self-reliant. In AD 406 much of the Roman army was withdrawn from Britain to deal with an invasion force of Vandals, Alans and Suevi into Gaul from across the Rhine. In AD 409 there was a significant invasion in Britain of Saxons and in AD 410, the Emperor Honorius sent a message to the remaining Roman forces in Britain telling them that from that time onward they would be completely responsible for their own defence. As the Roman force weakened, invasions increased, disrupting the previous relative social and economic stability. Invaders included the Jutes from Jutland, the Angles from Angeln and the Saxons from Lower Saxony. The Essex coast was at the forefront of these skirmishes, especially due to its proximity to the old Roman capital of Britannia, Camulodonum. The Saxons settled in coastal areas of Essex as well as Suffolk, Kent and Norfolk. From AD 410-442 Britain was effectively independent of Rome and the Briton leader, Vortigern of Wales actively encouraged Saxon settlement in return for their aid in defending the Welsh against the Picts and Irish. However, as the number of Saxon settlers increased, resentment to 'the other' rose to such a level that civil conflict ensued between the Romano-Britons and the newly settled Saxons. In AD 442 the Romano British controlled towns of Norwich and Colchester were stormed. The Saxons soon consolidated Essex but the economy was in freefall. Markets practically ceased to operate, the pottery industry was destroyed and farms suffered, resulting in many Romano-Britons fleeing to Gaul.

By AD 603 the Anglo-Saxons, as they were called by then, had colonised the entire country and gained control over all parts of England.[28] Much land at this point belonged to tribal leaders or local kings. This was apparently true in Essex although many historians of the period, a period which is sometimes referred to as the 'Dark Ages' due to the lack of evidential records, are wary of readily accepting the existence of the so-called 'heptarchy'.[3] It is supposed that if it was a kingdom at all then Essex, along with Sussex, was a lesser power, merely a collection of Saxon tribal groups. The Anglo-Saxon historian F. M. Stenton has written that:

> No east Saxon king was of more than local importance; and although an early Saxon occupation of Essex is proved by place-names of a primitive type, no other part of south-eastern England has yielded so little archaeological evidence of its condition in the heathen age.[29]

Certainly there was an established genealogical line of Essex leaders, evident through the succession of their names: Aescwine to Sledda to Saebert to Sexred and then Saeward, Sigeberht the Little, Sigeberht the Good, Sigere, Sebbi, Sigeheard, Swaefred, Offa, Saelred, Swaefbert, Svvithred, Sigeric and then finally Sigered whose rank was reduced by Mercian overlords.[30] When Mercia was defeated by Egbert of Wessex circa AD 825 the 'sub-kingdom' of Essex was absorbed into Wessex. The leaders of Essex were not kings in the way that we think of them today. They were leader figures, established within a wider network of Anglo-Saxon sub-leaderships and were often subservient to an overlord, most probably the leader of Kent, East Anglia or Mercia. In terms of geographical area, the 'kings' of Essex took territories situated

---

[3] 'Heptarchy' refers to the idea of seven Saxon kingdoms, which eventually merged to form the Anglo-Saxon kingdom of England. The seven kingdoms were Northumbria, Mercia, Wessex, Essex, Kent, East Anglia and Sussex.

much further from the current boundaries of the county. Middlesex formed a province of the 'kingdom' and London was established as the capital.[31] Present day Hertfordshire as well as some of Kent for a time also formed part of the east Saxon 'kingdom'.

As with other coastal areas of Essex, the Saxons settled in the area of the Naze, valued for its productive soils and Walton was known as Aedulvesnase or 'Aedwulf's promontory'. Although it is not known for sure, Aedwulf was probably a 'thegn' who either owned or managed the land around the Naze.[4] 'Nase' or 'Naze' is thought to have been a contemporary word for promontory, developing to 'Naze' or 'nose'. The Soke of Aedulvesnase included land in Walton (including many hectares now lost to the sea), Kirby-Horlock and present day Thorpe-le-Soken, although none of those place names would have been used by contemporary Saxons, certainly not before the reign of Athelstan. The evolution of Walton's name from 'Aedulvesnase' in the first millennium AD to Walton-on-the-Naze is a subject that should be dealt with at this early point. Although I shall be referring to the term 'Aedulvesnase' throughout the book, other historians refer to the spelling of the early settlement of 'the Sokens' was in fact 'Edulvesnasae' or indeed 'Aedulvesnasa'. So often, names at this time in history were simply spoken and descended through generations by word of mouth. Early language in the vicinity of the Naze would have been similar to the Celtic tongues of the Cornish, Gaels and Welsh but the Roman occupation had significant impact in terms of the influence and dynamism of spoken language. The Anglo-Saxons brought with them language derived from Ingvaeonic west Germanic dialects, which had been transformed into Middle English by the Eleventh Century, influenced by Norman French, Breton,

---

[4] A 'thegn' was an Anglo-Saxon nobleman.

Flemish and the languages of other visitors and invaders. When the Normans catalogued the country under Domesday in 1087, they simply spelt the manor's name phonetically as 'Aedulvesnase' and this is the spelling made use of in this book.[32]

Linguistics as a study can be important in helping us to plot cultural as well as social changes over time, through looking at alterations in language and the development of words. In her book 'Aelduluesnasa: Kirby in History', Rosemary Pratt has suggested that Aelduluesnasa (or alternative spellings) became shortened over time, through 'well established linguistic conventions', to 'Altun's Naze'.[33] However, there are too many possibilities to outline exactly how this happened. Ken Palmer, using documents located at St Paul's Cathedral library in London, has noted that by 1154 the manor was called Edulvesnasa, in 1235 Eduluesnase and in 1255, Edolvenesse.[34] Again, the reasons for disparity must lay in scribes noting the name from a word of mouth reference differently over time. By 1185, St Paul's referred to the area as 'Walentonie' or 'Waltona' and by 1220, Waltuna.[35] It grew to Walton-le-Soken most probably upon the sale of the manors to the Barons Darcy in the sixteenth century and 'Walton-on-the-Naze' was used as early as 1787 although the term was not commonplace until the Victorian era.[36]

In his 'History of Essex' (1768) Philip Morant explained his theory for the origins of the name 'Walton'. He wrote that 'the wall thrown up on this shore is what gave the name to this town or village'.[37] This is a logical argument. The most prolific chronicler of Walton's history, Peter Boyden, agrees with the idea but also suggests that the syllable 'Wal-' refers to a settlement of 'Welsh'.[38] It is true that there may have been a number of Welsh, or at least native Briton settlers or immigrants, but almost certainly not fair reasoning for the sole origin of the settlement. Ken Palmer writes how, if this was true, there would be many more Weala-Tuns' along this stretch of coastline.[39] It is unlikely that the settlement of Britons would be isolated from similar communities. There are other possibilities for the source. For example, 'Wæl-tun' refers to 'slaughter'. [40] One other rationalisation has been suggested by Ken Palmer as originating from the Roman era. He has suggested that the local people in this area were 'foreigners' and were called such by the Romans – 'Wealh' means strangers or foreigners – and the connection remained long after the Romans departed. Yet another suggestion from Palmer is that, in 1329, when a whale was beached and the flesh and oil distributed to locals, the town remained to be known as 'whale town' or indeed 'Wallton'.[41] Even when looking further afield to other 'Waltons' – Walton-on-Thames, Walton (near Felixstowe), Walton-on-the-Hill, Walton on the Wolds, Walton near Warrington in Cheshire - and unfortunately, there appears to be little connection and so a definitive answer to the name's origins remains elusive. In the case of the Naze, I opt for a combination of Philip Morant's suggestion, taking the cliffs as the inspiration for the origins of the town's name, literally 'Wall-Town', along with Rosemary Pratt's phonetic evolution, the name stemming from its predecessor manor, Aedulvesnase.

The Dean and Chapter of Saint Paul's Cathedral owned the manor of Aedulvesnase for over five hundred years and it is therefore important to explain how the Church developed in Essex. Roman attempts to establish the Church amongst the ancient British failed unanimously. Paganism had formed an established cord throughout the land, with numerous sects and denominations in different areas.[5] In AD 597 however, St

---

[5] The leader of each tribal group believed themselves to be descended from an ancestral god, for most Anglo-Saxon 'kings'

Augustine carried out a mission to Britain. He travelled to the court of Aethelbert, King of Kent and overlord of the whole of the south of England, who declared his conversion soon after Augustine's arrival.[42] Saebert, King of Essex, and nephew of Aethelbert, was the next to accept Augustine's teaching and a bishopric was established in London (St Paul's). However, by AD 616, upon the death of Aethelbert, the number of Christians in Essex had dwindled and it was not until AD 662 that another King, Swithelm, accepted the teaching of the Church following a mission by Bishop, later Saint, Cedd. The contemporary chronicler Bede described how Cedd's missionary work gathered 'much Church, great Church, to the Lord'.[43]

Osyth, sometimes known as Osgyth was another important figure in the development of the Church in Essex. Born to pagan, noble Mercian parents she was raised in a nunnery but was forced in to a dynastic marriage with King Sighere of Essex, apparently disabling her Christian aspirations. Nonetheless she was eventually able to establish her priory at Chiche, just a few miles from Aedulvesnase. Chiche or 'Chich' was the Saxon word for 'bend', relating to the 'bend' in Alresford creek which runs down to the Colne estuary. There is a story that at some point in the later part of the seventh century, Viking raiders stormed Chiche, destroying everything in their path. The nuns stood firm and Osyth stood up to the Viking 'chief', rejecting his terms that the nuns of the Abbey renounce their beliefs. The 'chief' was furious and ordered one of his men to behead Osyth on the spot which was duly done. Unfortunately, as Professor Ronald Hutton has discussed with me, the story is taken out of any historical context since there are not any records of raids having taken place in the seventh

this was from Woden but in the case of Essex, from Saxnot or Saxneat.

century. It was difficult for the Church to convince many in Essex to convert, as they had been practising pagan tradition for generations. Nonetheless, despite the volatility of this early period the influence of the Church in the county strengthened. Up to circa AD 850, the Diocese of Essex included minoters at Chiche, Bradwell, Wakering, Tilbury, Barking, Upminster, Bishop's Stortford and Hadstock. The period should be seen as one in which different sects of paganism and early forms of Christianity co-existed and collided in almost completely random patterns.

Christianity was consolidated during the Saxon leadership of Offa, who triumphed during the civil war of AD 757. He governed the people of England in a new way, treating the lesser kingdoms, such as Essex, as mere provinces of Mercia, granting them even less autonomy. Whether or not Essex was an independent province or 'kingdom', in AD 825 it lost any autonomy it had held upon Egbert's conquest. The Anglo-Saxon Chronicle, written during contemporary Saxon Britain, lays out how, in AD 825, Egbert, King of the West Saxons (Wessex) fought and defeated Beornwulf, king of the Mercians, at Ellandun and then sent detachments in to Kent defeating its king, Baldred which also resulted in the submission of the South and East Saxons (Sussex and Essex).[44] The East Angles turned to Egbert to protect them from the Mercians resulting in Wessex becoming all powerful. Egbert died in AD 839, his son Aethelwulf becoming leader of Wessex with his younger son Athelstan ruling Essex, Sussex, Kent and Surrey.[45]

Although the Saxons were slowly embarking on consolidating English territory under one system of rule, Danish raids continued in the ninth century. Aedulvesnase was particularly prone to these attacks.[6] The east coast raids took place through the eighth Century but they became so pugnacious by the ninth that the Danes had to be financially placated to avoid further damage to eastern territories. By circa AD 865, a time when Essex formed part of the Kingdom of Wessex, raiders increasingly brought with them settlers, who at first made their homes in Suffolk and Norfolk but then looked to expand. In AD 867, the Danes had captured York, the capital of Northumbria and inserted a puppet leader. Three years later in AD 870, East Anglia also formally fell under Danish rule. In AD 871 Alfred, often viewed as the legendary 'saviour' of the English people, was crowned as King of the West Saxons. However, by AD 875 the Danes had further consolidated their territory taking, Mercia and Essex. It was not until a treaty in AD 884 that relative peace was achieved and Danish rule was allowed within certain areas of England. This included territory in Essex and, for a generation, the people of Aedulvesnase were under the jurisdiction of the Danelaw.

In AD 911 the Saxons inflicted defeat on the Danes at Tettenhall, in modern day Wolverhampton, which resulted in both Essex and East Anglia becoming protectorates of Wessex with King Edward becoming their suzerain overlord.[46] By the

---

[6] The term 'Danish' is used distinctly from the term 'Viking'. 'Viking' refers to the more general Norse or Scandinavian movement whereby explorers, warriors, merchants and pirates from Denmark, Norway, Sweden and parts of modern-day Finland and Germany, raided, traded, explored and settled in expansive areas of Europe. The eighth to the eleventh centuries is known loosely as the 'Viking Age' when 'Vikings' travelled from Constantinople in the east to Newfoundland in the west and Greenland in the north. Raids on Aedulvesnase were however specifically and almost unanimously of Danish origin.

tenth century Essex and East Anglia had been liminal cultural spaces for generations, influenced both by Saxons and Danes. The Danes continued to push for influence in England however and the tenth century was riddled with attempts by Scandinavians to resettle in southern Britain. This was shown most vehemently at the Battle of Maldon on August 11th AD 991 although it should be said that not all historians regard the Vikings in a sole capacity as 'violent invaders'. They were also peaceful traders and settlers.[47]

AD 938 is prominent in the Naze's history. It is the year that the Dean and Chapter of St Paul's Cathedral were supposedly granted a charter by King Athelstan (AD 924-939) confirming the ownership of Aedulvesnase, along with various other lands and territories in Essex, Hertfordshire and Middlesex. However, the original charter to confirm the legitimate gift is not in existence today. The document that is available was fabricated during the early twelfth century to legitimise the Dean and Chapter's authority.[7] Certainly by the time the Domesday Book was written in 1086, the Dean and Chapter's authority in Aedulvesnase was undisputed, although the legitimacy of that authority can be disputed. Why would Athelstan have made such a generous decision as to grant these specific lands to the Cathedral? The answer may lie in two areas, those being the growth in the importance and influence of the Church in Essex and secondly, Danish raids, which meant that holding land on the Essex coast was expensive. Could Athelstan have been disposing of an unwanted burden? Perhaps a charter had been drawn up in AD 938 to confirm ownership but had been lost at the time of

---

[7] In the 'forged' charter it is stated that Athelstan, King of the English, confirmed that the Chapter of St Paul's held certain estates in Hertfordshire, Middlesex and, in Essex, '12 hides at Belchamp St Paul's and Wickham St Paul's, 8 at Heybridge, 30 at Edulvesnaesa and 12 at Runwell.'

the Norman Conquest or before. Alternatively the transition of ownership could have been a ceremonial royal gift to the Cathedral without written documentation of the deed, although this is unlikely. John Button has called the charter drawn up in the twelfth century an 'authentic fabrication', indeed a forgery but one that was recording 'a real bona fide royal gift' from the tenth century.[48]

The Church's influence continued to grow during the early Middle Ages and the ownership of land was a symbol of power that brought with it greater prestige. By controlling land the Church was able to act as a major economic player and as a result wield social power, converting the pagans. The Cathedral Church of St Paul was situated in the centre of the East Saxon metropolis of London and was for many, the central meeting place for all who lived in the city. It was founded by King Ethelbert of Kent in AD 604 as a Dean and Chapter foundation and the Episcopal seat of Bishop Mellitus. It is because of its geographical position at the heart of London that the Cathedral grew so rapidly in importance. It was the central church of the metropolis and grew to have many significant benefactors, including the Saxon 'kings'. Much of the land granted to St Paul's was gifted by the 'kings'. However, many of the documents confirming ownership were fabricated at a later date.

Other than during the period of Danish rule, Scandinavian raids were common until just prior to the Norman Conquest of 1066 when, in September, Harald Hardrada of Norway tried to take the country once and for all but was unsuccessful at Fulford and Stamford Bridge. In AD 991, the Danes inflicted a significant defeat on the East Saxons at Maldon, crossing the marshes from Northey Island.[49] The incident, which was fairly insignificant both militarily and politically but is seen as a vitally important and vibrant part of early Essex history, probably due to the existence of an epic poem, written by a contemporary minstrel, which survived through the ages, therefore embedding the event in to Essex cultural memory. A short section describing the midst of the battle is printed below:

> So stood firm the stout-hearted
> Warriors in the war - they did keenly strive
> Who with his point first should be able
> From fey men to win life.
> Warriors with weapons: wrack fell on earth.
> They stood steadfast; Brithnoth stirred them,
> Bade each of his men intend to the strife
> That would from the Danes win glory.[50]

The generation that lived during the turn of the second millennium AD suffered considerably for both political and environmental reasons. In AD 994, Essex once again suffered Scandinavian raids when Olaf Tryggvason and Swein Forkbeard, King of Denmark, forced the Saxons to pay the equivalent of £16,000 in blackmail payments, crippling the local economy for years.[51] 1014 was the first recorded incident of great floods across the East coast, important to any history of the Naze. The Anglo-Saxon Chronicle states that '...on St Michael's eve (28th September), the swollen incoming tide swept far and wide through many places in this land; and it ran further in land than it had ever done so before, and submerged many homesteads and drowned a countless number of human beings'.[52] The vast low lying marshes of Aedulvesnase were particularly prone to these large scale floods.

In 1018, upon the death of King Edmund, England (as it was at this point) was united under a Danish king, Cnut. Cnut was King of Denmark, Norway, England and parts of Sweden until his death in 1035. In fact, Danish rule in England continued until the crowning of Edward the Confessor in 1042.[53] The years of Danish rule instilled a greater peace within Essex and in England as a whole. It was the first time for decades when Scandinavian raids were not happening. In AD 962, the Danes had burned St Paul's Cathedral to the ground, but upon its new foundation and their subsequent invasion in 1018, they respected it. Cnut confirmed all the lands of the church, and intimated to his bishops, earls, peers and ministers that the priests of St Paul's were under his protection and their lands were free from burdens. This of course included the lands at Aedulvesnase and despite the lack of evidence regarding the Athelstan charter the Cnut proclamation gave St Paul's renewed authority over their land.

Why did the Danes embrace the institution of St Paul's so warmly in 1018 when fifty six years previously the Vikings had burned the Cathedral to the ground (possibly burning the Athelstan charter with it)? The most logical answer is that the Danes had managed to take the English crown and with it the control over the appointment of Bishops for St Paul's. A stone has been found in the churchyard of the Cathedral which bears the Runic inscription; *'Finna laet laeggia sten paenssi auk Toki'* or in literal English, 'Finna caused this stone to be laid over Tuki', which leads one to believe that the Danes were active in the Church during their time in power.[54] / [8] During this period Danes settled in the area of Aedulvesnase. The neighbouring villages to Walton - Kirby and Thorpe - have names of Danish origin and in the 1222 'Domesday of St Paul's' there is evidence of an Olaf in Walton and a Swe(i)n in Kirby, noted amongst the villeins, perhaps descendants of Danish settlers from the time of Cnut's dynastic period.[55]

---

[8] The stone is preserved in the Cathedral library.

In AD 1046, when Edward the Confessor sat on the English throne, Essex experienced both a minor Danish naval invasion, led by the outlawed Danish pirate Osgood Clapa, and a terribly severe winter. The Anglo-Saxon Chronicle describes it as having been '...so hard a winter...both the pestilence and murrain, and birds and fish perished through the great cold and hunger.'[56] By the time of the Norman Conquest, people in Aedulvesnase faced an unreliable life in the fields, governed by the nature of the weather and made more volatile by threats from Scandinavian skirmishes. They juggled the unreliability of subsistence farming and the weather with the volatility of international politics and warfare. The lordship of St Paul's Cathedral, beginning in the later tenth century, laid the way for land use to become more structured and regulated, as the parish was viewed increasingly as a productive estate for the Church, rather than a subsistence asset for those who lived there. In the next chapter the extent to which the Cathedral was involved in 'the Sokens' will be explained as well as answering the question of how 'the Sokens' came in to being in the first place.

CHAPTER 3

Walton and 'the Sokens' 1066 - 1544

The tenth century Benedictine monk Aelfric is famously attributed to the assertion that medieval society to be one of three orders - those who work, those who fight and those who pray. Although academic research has shown medieval society to be far more complex than this it remains one way of viewing and understanding it. Rather than being stratified exclusively in terms of classes or orders, medieval English society was made up of numerous different axes of social inequality.[57] Social identity was driven by class, order, gender, status of freedom, geography, economic activity, religious identity, profession and other forms. This resulted in multiple identities and various similarities and differences between individuals on a social basis. Nonetheless, no matter the position one held in life it was God who united all eventually. In medieval English society this resulted in great benefits for the established Church. For the clergy access to the corporate wealth of the Church (both its temporal wealth as a landowner and income from spiritual services drawn from particular ecclesiastical offices) was the consequence of their membership of a status group that was not based on the production of goods.[58] The clergy and aristocracy, with some key exceptions such as many Cistercian monks and members of some other monastic orders, shared an identity devoid of manual labour, fundamentally separating them from the lower orders and merchant classes. This particular identity brought with it considerable power and influence.

The Church was a substantial landowner. Walton-le-Soken formed a part of the estates owned by St Paul's Cathedral known broadly as the 'liberty of the Sokens'. The Soken designation coupled Walton-le-Soken with Kirby-le-Soken, Horlock and Thorpe-le-Soken and governed the specific privileges to which people were entitled.[9] It was part of a specially warranted area – at its simplest Walton 'in' the Soken.[59] The word 'soke' or 'soken' was given by the Anglo-Saxons to areas that hold special jurisdiction. The Reverend Philip Morant wrote, 'Soken is derived from the Saxon *Soc* or *Soca*, signifying a peculiar power, authority, or liberty, to administer justice and execute laws within itself, and likewise the circuit or territory wherein such power is exercised.'[60] Rosemary Pratt, in her 1976 publication about Kirby-le-Soken, debated this definition, writing that 'the word does not simply mean 'peculiar jurisdiction', as there were dozens of 'peculiars' that were not Sokes'.[61] 'The Soke', she argued, was an estate different to a manor in three principal aspects. Firstly, it was a scattering of different holdings or villages, not just one. Secondly, a free Sokeman could not give allegiance to a lord outside the soke, unlike other freemen who had this ability and thirdly, a Sokeman was not obliged to give the lord a set number of days of service, unlike other freemen.[62] Clearly, the inhabitants of a soke enjoyed rather different privileges from those living outside its borders. For example, nobody resident in the Sokens could be arrested by the royal sheriff without a representative of the lord of the manor being present.[63]

Further, the powers and authority of the King's sheriff were limited within the jurisdiction. As discussed in the previous chapter, the manors of the Sokens were granted to the Dean

---

[9] Little is known about the settlement of Horlock although for geographical reasoning I have referred to it as a subsidiary of Kirby.

and Chapter of St Paul's Cathedral, London by King Athelstan (AD 924-39) in the tenth century although there is no legitimate charter to confirm this acquisition. It is clear that by the twelfth century, St Paul's had been granted a significant area of land across the country, but particularly in Essex and along the coastal fringe. In 1181, during a particularly cold winter, the Dean of St Paul's suffered 'considerable discomfort' during a week-long inspection of Essex churches, travelling from Belchamp to Wickham, to Kirby-le-Soken, and then to Heybridge, Tillingham, Runwell, Barling, Norton, Navestock and Chingford.[64] This list of parishes barely scrapes the surface of Essex manors controlled by the Cathedral.

The original Aedulvesnase estate consisted of thirty hides,[65] usually regarded as roughly equal to around a hundred and twenty acres each, the notional amount of land able to be ploughed by a standard team of oxen in one year.[66] Although that is a good starting point, in reality the area of a hide was far more variable. In Anglo-Saxon England, a hide had been a way of measuring the taxable value of an area of land. The size of each individual hide was therefore dependent upon several factors. It was the quality of the land that was important, rather than whether the land was of arable, forestry or pastoral use. Ploughing rates depended on both the terrain and the workable nature of the soil.[67] Either way, by the eleventh century, the estate at Walton had been reduced in size from thirty to twenty seven hides, due to the creation of a new three hide manor in Kirby-le-Soken known as Birch Hall.

Birch Hall is mentioned in the Domesday Book as 'Birichou' or 'Birch Hall with Horsey':

> Birichou is held of the Earl (meaning Eustace, Earl of Bologne) by Robert (the tenant): it was held by Ingelric of St Paul, London, for one manor and for three hides.

Then six bordars, now eight. Then two serfs, now one. Then in the demesne two teams, now none. Then among the homagers one team, now two. Wood for ten swine. Pasture for one hundred sheep. Then two horses, now none. Then eight beasts, now none. Then eight sheep, now none. Then six swine, now none. Then worth sixty shillings, now four pounds and seven shillings.[68]

After a long period of separation Birch Hall was granted to the Priory at Chiche (now St Osyth) and reunited with the Sokens upon the acquisition of the Priory estates by the Barony of Darcy, following the Dissolution. When I visited Birch Hall to interview John Fleming during the period of researching for this book I couldn't help but be struck by the acutely separate identity of the place. The farmstead stands on much higher ground than the surrounding area, overlooking the Walton Backwaters, Harwich and Felixstowe to the north, Kirby and Walton in the distance to the east and rural Tendring countryside to the south and west. Birch Hall has a fascinating history, not to be relayed here, although some of it is touched upon in more detail in the oral history interview with John Fleming, set out in the appendix. Prominent in its more recent history however is marked by the national monument situated near the sea wall to the north of the farmstead. Following our discussion John drove me down in the land rover to take a look at it. During the Second World War decoy lights were lit on the saltings in an attempt to mimic Harwich and encourage German bombers to drop their loads on the mud rather than the port. A similar decoy was found on the northernmost edge of the Naze. The site of the Birch Hall monument consists of a number of structures. Firstly a substantial brick built shelter with a thick concrete roof and blast wall. There were two rooms, one holding a generator and the other living quarters for two men. The generator was used to ignite the oil lights on the saltings, the oil being pumped from a nearby tank. The structure is now preserved as a national monument, testimony to a small part of Birch Hall's diverse history.

It has been argued that 'Horlock', the fourth constituent settlement of Aedulvesnase was part of Birch Hall manor. Ken Palmer has argued that the 'mysterious settlement' can be explained as the separate manor of 'Birch Hall and Horsey'.[69] 'Horsloc' is the Middle English word for 'horse enclosure', supposedly Horsey's original land use. Palmer suggests that 'Horslock' became 'Horlock' by the slip of a monk's quill.[70] However, although this is an inspired suggestion, it must be ruled out since Horlock/Orlock/Orlok formed a core part of the Aedulvesnase manorial complex held by the Dean and Chapter of St Paul's, evident in documents dated 1150, 1222 and 1297, not a part of Birch Hall. Horlock may have been located further east, perhaps in the areas known today as New House farm or Hedge End Island. Without a thorough marine archaeological expedition it is impossible to know where Horlock was located, who lived there and why.

One of the most significant events in eleventh century England, other than the Conquest of 1066, was that of the compilation of the Domesday Book in 1086. The years directly preceding Domesday were difficult for North Sea facing manors. In 1077 there was a particularly hot and dry summer that resulted in widespread fires across the countryside and the Essex landscape was particularly affected.[71] Most local houses and churches were made of wood and burned to the ground, and wild fires spread in forests and arable lands. In 1085, the Sokens faced similar prospects as William I initiated his 'scorched earth' policy, wasting coastal Essex communities to deter a possible projected invasion by Canute IV of Denmark.[72] The Conqueror had feared the threats of Sweyn II of Denmark for years and sacrificed productive parts of his kingdom for the sake of defending the remainder.[73] This was followed in 1086 by a year of 'extreme pestilence', in which significant numbers of the population starved.[74] It was these complications and threats from abroad that led to the royal commission of a complete survey of everything in the land, the very first of its kind: the Domesday Book.

Domesday was the first time that Essex was described as a separate county. The national project was divided in to two key works. The 'Little Domesday book' compiled details of the counties of Essex, Suffolk and Norfolk. In the account, St Paul's justified their claim to the Sokens, lacking written evidence to prove the legitimacy of their holdings. St Paul's copied the Domesday account of Aedulvesnase verbatim into their own 1297 survey to give the assertion that the grant of the thirty hides by Athelstan to St Paul's had absolute legality, when actually no charter or title could be produced.[75] Domesday was both a symbol and a signal of Norman power. By scribing land ownership in a written document the Normans legitimised, for themselves, their control of the land and granting of estates to tenants of the Crown. The whole purpose of Domesday, in addition to acting as an early detailed census, was to legalise, in perpetuity, the seizure of English land by the followers of William of Normandy. St Pauls' claims to Aedulvesnase became fully and officially validated. Whereas many estates controlled by nobles were transferred directly from Saxon to Norman ownership upon the Conquest, lands belonging to the Church, almost exclusively, remained in that specific ownership afterwards. Different political factions may have been in charge of the institutions but the institutions themselves prevailed. The officers of the Church were, overwhelmingly, newly patronised Normans, similar to the period of kingship of Cnut, who patronised Danes to positions of power in the English Church.

45

In Domesday the three Soken parishes were treated as a single unit (Aedulvesnase) with the following stated:

> Aedulvesnase has always been held by Saint Paul for a Manor and twenty seven hides. There were then (meaning at the time of Edward the Confessor) eighty six villeins, now sixty three. There were forty bordars, now fifty. Then as now six and six ploughs in demesne. Among the men then sixty ploughs, now thirty. There is woodland for three hundred pigs. There is nine acres of meadow. There are now two mills. There were then three saltpans, now two. There is pasture for three hundred sheep. There are twenty two head of cattle, thirty pigs, two hundred sheep and four hives of bees. It was then worth £26, now £30 and 1 mark of silver.[76]

The Aedulvesnase extract in Domesday is of interest for several reasons. Firstly, we can see both a decline in population within the district and the reduction of plough teams from sixty to just thirty, implying a substantial decline in the amount of time that villeins[10] were able to work on their own strips, in favour of a more substantive communal effort on the demesne. John Button has suggested that this could have been when the people of Walton lost their status as 'hidarii'[11], which they had once shared with the other constituents of the manor (Kirby, Horlock, Thorpe). Individual villein holdings were dramatically reduced from 17.5 acres to just 5 acres, along with an increased amount of labour demand on the demesne. This suggests that the coming of the Normans

---

[10] A *villein* was a member of a class of partially free people under the feudal system, who were serfs with respect to their lord but had the rights of freemen with respect to others.
[11] *Hidarii* refers to tenants on land assessed in hides. *Hidarii* had various privileges above those of other *villeins*.

instigated a ruthless alteration in social status for individuals as well as in the organisation of physical labour collectively.

The Sokens were eventually called by contemporary locals by their constituent parts, Thorpe, Walton, Kirby (and Horlock) Walton was the principal settlement of the manor, due to the status of the demesne and curia[12] buildings. In Medieval England, the demesne was the part of the estate under the direct control of the lord of the manor, or in this case the individual who held the lease from the Dean and Chapter of St Paul's. This individual would have been a high ranking cleric or official of the Cathedral. In 1222 it was William de Burnham and in 1292, Canon Fulk Lovel, Archdeacon of Colchester.[77] The lord of a medieval manor controlled not only his estate but his people. It was a small dominion, held courtesy of the Crown, within which the lord was emperor over subjects of different ranks, his power only limited by laws and customs of the Crown. In medieval society status was a key measure of wealth and power that both enabled influence. The lord of the manor had ultimate power over criminal justice and the 'Court of the Sokens' existed at Kirby Hall for many centuries after the departure of the Cathedral from the area, with gallows standing at Newgate Street in Walton.[78] Lords treated their manors as small principalities over which they were sovereign. The demesne acted as a 'home farm', with the lord delegating day to day responsibilities to a steward. The remainder of the manor was allocated to villein 'subjects'.

The demesne consisted of arable lands, upland, meadows, woodland and marshland for sheep, hunting, wildfowling and fishing. At the centre of the demesne was the manor, consisting of a hall house with various farm buildings. The original medieval site was further east of the current Hall and

---

[12] A *Curia* refers to the principal buildings of the manorial complex, surrounding the manor house.

farm buildings, being rebuilt further inland as the sea encroached upon the site. The remainder of the lands of the manor were allocated to villeins or tenant farmers. The specific aspects of the manor buildings will be discussed in a later chapter. The management of Aedulvesnase differed to the Cathedral's other estates, in that each hide was treated differently and held by tenants in different proportions. Equality was by no means the rule and often the lord of the manor favoured certain tenants, called Hidarii, over others. For example, in the 1222 survey known as the 'Domesday of St Paul's', the first hide was shared between four tenants, whereas the second was shared between nine, the third among ten and the fourth amongst eight.[79] Some tenants had better reputations than others and consequently received greater rewards and liberties.

Thorpe, Kirby and Kirby-Horlock appear to have been organised on a different basis to land holdings in Walton. Holdings in Thorpe, Kirby and Kirby-Horlock averaged at fifteen acres with a prominent aspect of 'freeholder tenants', whereas Walton consisted of smaller holdings with a more intensive labour service system.[80] This denotes one of the peculiar differences between Walton on the one side and Kirby, Horlock and Thorpe on the other. In 1297, Walton was characterised by small villein holdings of around five acres as well as heavy labour service requirements. In the other hidal vills[13] this was replaced with a system of rents. By the thirteenth century the demesne at Walton (as well as those who lived and worked the land there) concentrated on food production, with people in Thorpe, Kirby and Horlock left to raise monetary funds for the Cathedral through rents and taxes.[81] Although villeins of the Soke held special privileges, it

---

[13] *Hidal vill* refers to any settlement governed by the hidal system.

would be unwise to assume that the tenants of Aedulvesnase enjoyed benefits above and beyond the villeins of other St Paul's estates. Duties similar to those carried out on other manors had to be performed, although some tenants may have been trusted with greater responsibility. Most of the Cathedral's lands provided the Dean and Chapter with weekly renders of food, known as firmae. However, due to the distance of the Sokens from London, Walton-le-Soken manor was used to feed the local population. Conjecturally, it may be assumed that any excess produce was sent by boat from the Naze to the Cathedral warehouses in London but critically there is no documentary evidence of the estate providing weekly firmae.

Before the Normans came to England, there was a second, smaller manor at Walton which had been endowed by a prebendary of the Cathedral. Robert Losinga, Bishop of Hereford was the first known holder of the position although it is probable it was not he who endowed the installment.[82] By 1327, the manor was simply referred to as 'Consumpta per Mare' as it had long been lost to the sea, possibly since the 'great floods' of 1014.[14] Since then Walton Hall remained the principal manor for the entirety of what is now the modern day parish of Walton. In 1222 the demesne consisted of 750 acres of land with a further 466 acres tenanted and extensive marshland pastures used by all tenants within the liberty of the Sokens.[83] In 1297, the demesne had grown to 819 acres with an 'unmeasured amount of marshland'.[84] However, in later times land use altered, shifting from arable to grazing. In 1676 just 360 acres of land were recorded with 1000 acres of salt marsh, to be used as grazing marsh.[85] Although the land

---

[14] This manor continued to be recognised as a prebendary stall in the Cathedral. Notable holders of the prebendary whilst it was 'on land' were William Guffard, Bishop of Winchester (died 1128), Turstin, Archbishop of York (died 1140) and Richard, Archbishop of Canterbury (died 1183

area for potential farming exploitation increased with the reclamation experiments carried out from the Tudor period onward, so that today the area of enclosed land is greater than the 1676 figures, the diminution of the area illustrates the rapidity of land loss at the Naze. The focus today may be placed principally on the erosion of the cliffs but five to eight hundred years ago the visual emphasis of coastal risk was placed on the usable marshland, which people relied upon for grazing. This low lying area was little protected and was therefore at the greatest risk every time a surge situation took place.

The lease of the manor was almost exclusively granted to clerics but sometimes to laymen who were closely involved with Church affairs. The Dean and Chapter of the Cathedral tried to prevent any tenant establishing a hereditary line and therefore a claim on the land, and so clerics were overwhelmingly the preferred guardians of the Soken manors. In AD 1297 Canon Ralph de Baldock, Dean of the Chapter of St Paul's, commissioned a thorough survey of the manor of Aedulvesnase.[86] There were seventy one households in Walton alone, all of whom provided services to the lord and provided him with a share of the produce of their tenanted acreage. The lord required each tenant to complete one 'job' per week on the demesne for every five acres of land that each villein held. John Button has noted that this could include the threshing of three bushels of wheat, or three bushels of beans, rye or peas, or half a quarter of barley and six bushels of oats.[87] However, although villeins had to perform similar duties, people within the Soke had several unusual privileges that set them apart from villeins outside of the area. In 1509 these customs and privileges were formally established.[88] Villeins of the Soken manors were entitled to sub-let their land for a period of up to fifty years without the need of seeking permission from the lord. They were allowed to hunt and hawk for game and to catch fish on their holdings without any external interference, unthinkable elsewhere.[89] Furthermore, the land holdings of widows were safeguarded, going much further than the limited number of provisions that widows were entitled under Magna Carta.

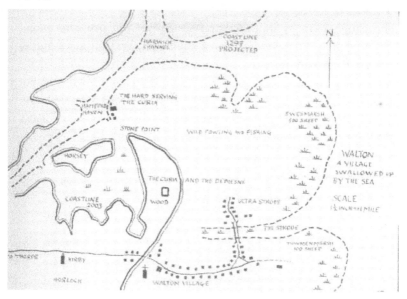

Above: John Button's drawing displaying his projected coastline in 1297 (dotted line) compared to 2003 (solid line). Courtesy of John Button.

The 1297 survey sets out the exact formulation of buildings and land as they existed in that year. The demesne contained four carucates of arable land. A carucate, literally 'plough' or 'plough team' in Latin, was a medieval unit which consisted of a hundred and eighty acres split in the three field crop rotation system practised at the time. Sixty acres were planted in the winter, sixty in the spring and the remaining left to fallow for a year. Peter Edwards has argued that acres were not precise measurements until statute measurements were established in the thirteenth century. An acre was merely one or two strips of land with small strips counted as 'half acres'.[90] Evidence from the survey also suggests that the villeins who lived at Walton-ultra-Strode (beyond the Strode), an area often cut off from the main village by the sea, had the right to graze their oxen on the lord's fallow for a small fee.[91]

The rough map by John Button shown on the previous page explores how the geography of medieval Walton may have looked. It should be noted that until a thorough archaeological survey takes place, both on land and out to sea, it will not be known where the exact positioning of the curia and house have been over time. John Button has made this point very clear, although in his assessment of Walton in 1297 he places the curia at the site where the current farm buildings are.[92]

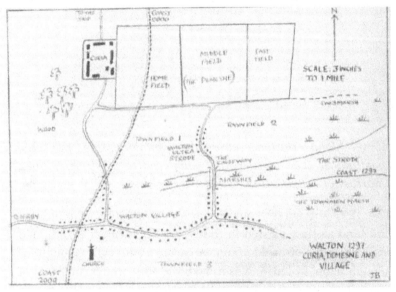

Above: Sketch of the estimated coastline of Walton and possible positioning of the medieval curia in 1297 (opposite). Sketch of Walton's possible geography and field layout in 1297 (above). Courtesy of John Button.

Walton-ultra-Strode was divided from the main village by a stretch of tidal mud and saltings known as the Strode. The lord allowed the people of Walton-ultra-Strode to graze oxen on the fallow and so it is assumed that this part of the settlement was conveniently situated nearby to the curia and demesne. The Home Field or 'Holmfeld' was situated nearest the curia and consisted of 164 ½ acres, its annual rentable value being £4-2s-3d. Next to it lay the Middle field or 'Middelfeld', consisting of 218 ½ acres and worth £4-19s-6d. The Errefield, later East field, was the 'ploughed field', consisting of 204 ½ acres and being £4-12s-4d in monetary worth.[93] The word 'ere' means 'to plough' in middle English. There were also two much smaller fields, the Bancroft de Assieslond or 'Bean croft on the demesne', forty two acres in area and Overdoune and Chalfpitel or 'Overdown and Calfpightle', being just six acres.[94] Fields were divided in to strips in the pre-enclosure era and each individual received about half an acre of land to work. Meadowland consisted of five separate fields amounting to about fifteen and a half acres in total and there were a further seven fields of pasture, distinct from the considerable amount of marshland used for grazing sheep. These fields amounted to thirty three and a half acres and included the fields of Coulase, Batelond, Hardgrass, Longmead, Borsudoune (Boarsdown), Chalfpitel apud Banebos (Calfpightle by the Beans) and Wildmead. Additionally there was extensive marshland which included a hundred and sixty acres on Ewenemersch (Ewes Marsh) where four hundred sheep could be sustained.[95] Wildmersche (Wild Marsh) could sustain a hundred and fifty sheep and a further hundred sheep on Tunmenmersch (Townmen Marsh).[96] Grazing on this particular marsh was an established right as part of the set of privileges in being tied to the Sokens.[15]

---

[15] Several hundred years later, in 1900, the field names at

Throughout the medieval period, sheep from the Essex marshlands were valued not only for their mutton, skins and wool but also as a source of dairy produce, especially for cheese. In the Tillingham entry for the 1222 Domesday of St Paul's, 'marshland sheepwalks' are linked with their dairies (wiches or wicks): 'In the marsh are four sheepwalks, of which one is called Howich and can carry 180 head of sheep, another is called Middelwich, and can carry 130 head, the third is called Doddeswich and can carry 132 head, and the fourth is called Pirimers and can carry 110 head'.[97] The connection to this important long standing aspect of the local economy is remembered through place names such as Jaywick and Cockett Wick Farm in St Osyth.

Of the arable, each field was divided into long, narrow strips, or selions, the shape being dictated by medieval ploughing practice. Some of the area of these fields was allocated to tenants but the vast majority of customary tenants worked on strips in the vast open fields, separate from the demesne. These other fields are marked as 'town fields' on John Button's map but geographical positioning is unknown. Twenty eight acres of woodland are also mentioned on the demesne, of which

---

Walton were thus: Belgunies, Bury Field, Butts, Camping Field, Church Field, Clays, Cock's Land, Davis's, Drift, Elliotts, Floras Piece, Grove Field, Hall Hatch, Great and Little Harrods, Hicks Folly, Hog Island, Hungerdowns, Kitchen Field, Knowles, Lumber Leys, Moat Pightle, Mulberry Field, Poll Field, Pump Field, Rail Marsh, Roses, Round Ridges, Salting, Shallow Shot, Slipe, Spring Field, Stockings Pightle, Terrace Field, Upper and Lower Thistly, Threecorners, Town Field. Today, in 2015, the field names of Walton Hall farm are: Thistley, Mill field, Threecorners, New Zealand (where the AWA sewage works is situated), Wind field, Garden Field, Wood Field, Long Meadow, Hall Marsh, Foundry Marsh, Horse Marsh, Wood Marsh, Sheep Marsh, Great Marsh, Corner Marsh.

the customary tenants had the right to coppice seven acres of wood in any one year, leaving three quarters of the trees to maintain growth and therefore sustain the practice. It is likely that these trees were of varieties such as Hazel and Ash, perfect for coppicing, but it is possible that Oak was also planted and cut, using a longer rotation.[98] Wood was a vital part of the village economy, used for faggots and charcoal for fuel, for making sheep hurdles, construction and for making fish weirs. Outside of the field systems and marshlands, the Walton Backwaters fell under the domain of the lord. Wildfowling and fishing took place on a small scale. There would have been a very large number and variety of birds flocking to the Backwaters, as they do today, and this opportunity for extra food was exploited. Wildfowling nets were probably placed across the birds' flight paths so they could be caught at night. In the 1222 'Domesday of St Paul's' small fixed fish weirs are mentioned as having been used, costing tenants such as Ralph the Hayward and Thomas Battayle 1d per year.[99]

The trading, mining and fishing industries, so important in Walton's Roman and Anglo-Saxon past, became second to agriculture during the medieval period. Nonetheless, trade was important and food products were exported from Walton. John Button has inferred that there may have been a small quay or village hard present somewhere to the north of the Naze, where produce from the Great Barn of Walton was exported to other villages and towns further afield as well as to the London warehouses of St Pauls.[100] A thin channel continued north of Horsey to Landermere. Evidence of the harbour existing comes from an eleventh century tale of asylum. In 1052 the Archbishop of Canterbury fled to Normandy via Walton. He rode to Eadulf's Ness (Walton) with Ulf of Dorchester and from there took the one small fishing vessel available and sailed to Normandy.[101] Due to the nature of the vessel available we can infer that the harbour was fairly small.

During the medieval period the area of the Sokens was a productive agricultural estate, capable of sustaining a population which remained stable despite encountering invasion and disease. Despite the social hierarchical system of the time subjects of the lord of Aedulvesnase experienced special legal privileges, meaning they were more greatly protected both from labour exploitation and the Crown. However, people continued to suffer from severe weather events, famine and threats of invasion. From the sixteenth century major changes in landownership took place as across the country ecclesiastical estates were granted to members of the aristocracy. This narrative continued for hundreds of years before landownership became more 'local' in identity, with land often controlled by owner-occupiers. Despite the major change in landownership in the sixteenth century there does not appear to have been any key disruption as a result. It was the changes in central ecclesiastical policy, as well as pragmatic action taken by St Paul's Cathedral regarding their estate as a whole, that led to a new order at the Naze and the beginning of a slowly evolving privatization of the landscape.

CHAPTER 4

The Lie of the Land 1544-2015

Perhaps the most significant social change in sixteenth century England came as a result of Henry VIII's dissolution of the monasteries, a period of monastic suppression that lasted from 1536-1540. The dissolution took place as part of a transition of power from the Vatican to the Tudor court and followed the creation of the formal Protestant Church of England in 1534. The majority of monasteries were smaller abbeys or priories, such as the Augustinian priory at Chiche, known today as St Osyth. Monasteries had been closed before. Sequestration of monastic assets was a prominent policy of Edward III in the 14th Century. Cardinal Wolsey had, through the power of the papal authorities in Rome, closed many smaller 'decaying' monasteries in 1524 and 1527.[102] The funds of this action led to the expansion of his Grammar School in Ipswich – the King's School at Ipswich, now known as Ipswich School and also to the foundation of Cardinal College, Oxford, now known as Christ Church College.

However, Henry and his adviser for religious affairs, Thomas Cromwell, went much further than Wolsey in their attack on monastic control in the country. In March 1536, Parliament passed an Act, whereby monasteries earning less than £200 a year were formally dissolved, with their property passing to the Crown.[103] About three hundred religious institutions fell in to this category and all but sixty-seven were dissolved and subsequently dismantled.[104] The Priory of Chiche (St Osyth) was taken over by Henry in 1539, and in 1540 he granted it, along with the priory lands, to Cromwell, who he also made the first Earl of Essex. However, following Cromwell's execution soon after in July 1540, it was gifted to Henry's eldest daughter, Mary. She held the priory until Edward VI granted it, in 1551, to its then custodian, Sir Thomas Darcy, who became Lord Darcy of Chiche. [105] Many of the buildings of the medieval priory, including the great Abbey Church, were destroyed in the Dissolution. In 1553, upon being granted full ownership of St Osyth Priory, Darcy undertook major renovations and expanded the building extensively, adding two towers. He attempted to prove to Mary his loyalty by helping to quell the spread of William Wyatt's rebellion in Essex and also by supporting the restoration of Catholicism in the county.

In 1544 it was agreed that the estates of the Sokens, which then comprised Walton Hall, Thorpe Hall and Kirby Hall, would be exchanged by St Paul's with the Crown for more profitable estates closer to the capital. Quite why the Crown agreed to the transfer is unknown but nonetheless the decision marked an end of nearly six hundred years of riparian history for the Cathedral in the Sokens. In 1551 all of the Soken estates, along with other lands in the Tendring Hundred, were granted to Sir Thomas Darcy.[106] Darcy was a prominent member at the courts of both Henry VIII and Edward VI. He managed to push himself forward as a beneficiary following the dismantling of the St Paul's Cathedral estate. As well as the Soken manors Darcy took land at Wivenhoe, Tendring and Chiche (St Osyth).

Sir Thomas was born on 4th December 1506 in Essex and was the only son and heir of Roger Darcy of Danbury and Elizabeth Wentworth, daughter of Sir Henry Wentworth of Nettlestead, Suffolk.[107] His father was a considerable landowner both in Essex and Suffolk, and Thomas was brought up knowing that, as an only child, he was the sole heir of the family estates. He was also the heir to the estate of his great-uncle Robert Darcy and would further inherit his mother's property upon her death. Thomas's parents died before he came of age and so he was put under the wardship of Sir John Rainsford of Bradfield, Essex, the father of his future wife. Thomas was first married, aged just fourteen, to Sir John's daughter, Audrey, in September 1521. However, the marriage, taking place at what we see now as a very young age, did not work for each party and so he was married secondly in 1532 to Elizabeth, daughter of John De Vere, 15th Earl of Oxford. The De Veres were a prominent Essex family, heavily integrated in the high ranks of the army, and so Thomas found he could increase his stature as an officer.

When the 15th Earl died, in March 1539, Sir Thomas was granted the three North Essex titles of the De Vere family – the stewardship of St Osyth Priory and Tendring Hundred as well as being made Keeper of Colchester Castle. However, the Naze and the Sokens remained the property of the Cathedral throughout this time, still tied to the Dean and Chapter. Sir Thomas was trying to establish himself as a court adviser but experienced spectacularly slow progress in his attempts to get to the centre of power. He began as a member of the household of the King's illegitimate son Richmond and therefore could access the central court. However, he was not recognised as a future prominent courtier at this point and was unable to make many steps towards doing so, lacking the contacts necessary to introduce him to 'the system'.

His luck changed when he met the Duke of Somerset, Edward Seymour, with whom he was to become a great friend. Seymour's sister Jane, upon the execution of Anne Boleyn in 1536, became Queen Consort of England and Darcy's friendship with Somerset meant he had a far greater chance of increasing his influence at court. He became Master Armourer to the King in 1544, as well as captain of the guard and he commanded the pensioner force in the expedition of 1544 against the French. However, in May of that year he returned to Tendring to oversee vital work on sea defences. Returning for this work was also an opportunity for him to personally survey his newly acquired Essex estates. He remained in Essex for a couple of months, residing at St Osyth Priory, and assisting the people of Essex in working to prevent an invasion by the French in that year. In June, he joined the Earls of Essex and Sussex in arranging for the defence of the Isle of Sheppey. Afterwards, he returned to court in London and was prevented from leaving as he was regarded as a key negotiator for the peace talks with France, when they were eventually held in 1546.

Darcy sat in at least two of Henry VIII's Parliaments. He was knighted at Calais in November 1532, when he may have taken up his seat in the Commons. Soon afterwards he received several leading posts in the royal household, including a Privy Councillorship and a knighthood of the Garter. As Lord Chamberlain, Darcy was one of the leading figures in the land during the final years of Edward VI's reign, and it was in this capacity that he signed the device enabling Lady Jane Grey to succeed to the throne and helped to proclaim her Queen. Northumberland asked Darcy to order Baron Rich to hold Essex against Mary, which he did, but on realising the popularity of Mary's cause in his county he forsook Jane and advised Northumberland to surrender. For Darcy's support of her rival, Mary dismissed him from office and placed him under house arrest at Wivenhoe. However, on 1st Nov 1553 he was pardoned through the intercession of his brother-in-law, the Earl of Oxford.

It was under Darcy that Walton Hall's identity shifted from a relatively important, if isolated, part of the St Paul's Cathedral estates to a small part of the landed 'empires' of the aristocracy, an identity which continued up until the late nineteenth century, when farms became more local in identity, managed by owner-occupier farmers. Walton had a wholly agrarian economy from the sixteenth to eighteenth centuries and was under the charge for many years of estate stewards, such as William Field in 1745 and Robert Lloyd before him, who resided at the Hall house. Field himself was part of the illegitimate ancestral line of Richard Savage, lord of the manor 1695-1712.[108] Before 'enclosure' (subdividing fields into smaller sections through the planting of many miles of hedges) the medieval open field system continued and stewards contented themselves in allowing tenants to 'farm in a husbandable manner' and to 'commit no waste'.[109] However, particularly in the early modern period (16th, 17th, 18th centuries) the instructions became more precise, with covenants emphasising the need for keeping the land in good heart by manuring or by forbidding overtillage.[110] This was a nationwide phenomenon, part of the great agricultural improvement movement and was true until full enclosure took place, early in the case of most Essex manors, and farms were rented as single area units.

The manors of the Sokens remained within the Barony of Darcy and descended through various Darcy sons, all of whom were normally non-resident in Essex, although the second Baron of Chiche lived at St Osyth Priory for the majority of his lifetime. The first Baron died in 1558, leaving the lordships to his son and heir John, who had been born at Chiche (St Osyth) circa 1539 and resided there his entire life. A proud Protestant, John was keen to ingratiate himself with the new Queen Elizabeth, who was crowned in 1558. He was knighted at the Coronation on the 15th January and took up his seat in the House of Lords soon after. Elizabeth wanted to express her strength in leadership as early as she could in her reign and undertook a lengthy tour of the aristocratic seats of England. St Osyth's Priory was one of the seats on the tour and in 1561 Sir John found himself preparing for a Royal visit by the entire Court of Elizabeth. St Osyth is renowned today for being one of the driest parts of the country but on the day of the Queen's arrival on the 30th July 1561, there was a storm, so ferocious that many thought the day of doom had come for the world. The Queen did not stay longer than a day.[111] However, she returned in 1579, staying for three days on a tour of Essex and Suffolk.

Sir John died on 3rd March 1580 and was buried at St Osyth Priory, where statues were erected in memory of him and his wife. In his will, to subvert the debts that would otherwise be placed upon his son and heir Thomas, he granted a lease of Walton Hall to the value of £406 13s 4d to Sir John Brocke, after which it was agreed that the estate would revert to his son, Thomas. Brocke died in 1586 when the estate came to Thomas who became the third Baron of Chiche, at the age of twenty one. Despite marrying twice, his only son and heir apparent, Thomas Darcy, died during his own lifetime, affronting Sir Thomas with the possibility of being the last Darcy to sit at Chiche. To mitigate the blow, on 8th October 1613 he declared his son in law, Sir Thomas Savage, his heir. Savage had married his eldest daughter Elizabeth and was made Viscount Savage in 1626.[112] The Savages were a wealthy family whose estates centred upon Cheshire and the north west of England. Sir Thomas Darcy was created Viscount Colchester in 1621 and soon after became Earl Rivers in 1626. With the succession in place Darcy felt confident that his Essex estates would not, at least directly, leave the family line. However, tragedy struck yet again, in the eyes of Earl Rivers, as Sir Thomas Savage died in 1635. This may have triggered the fissure in the health of the Earl himself who died soon after in 1640. The titles passed to Savage's son, John, and the

Soken estates to his wife, Elizabeth, Countess Rivers, the daughter of the Earl.[113]

Elizabeth was born at St Osyth Priory in 1581 and married Sir Thomas Savage on 14th May 1602.[114] They had eleven sons and eight daughters. The problem for the system of male line inheritance came upon Sir Thomas' death in 1635. Elizabeth, who was by now Viscountess Savage, was the only probable candidate for the inheritance of the Barony of Darcy. She inherited Melford Hall, which, along with St Osyth Priory formed her principal residence. As well as the Soken estates she also took the manor of Clacton, resulting in a significant estate. On the death of Lady Savage's father in 1640, the Earldom of Rivers passed not to her but to her eldest son, John. As compensation, she was created Countess Rivers for life in 1641. However, Lady Rivers faced a problem which married the personal with the political. She was staunchly Catholic and had served previously as a lady of the bedchamber of Queen Henrietta Maria, Queen Consort to Charles I. Sir Thomas Darcy, Elizabeth's father, had been excluded from becoming a county magistrate due to his strong Catholic allegiance. Elizabeth was never fully trusted by the people of St Osyth and suspicion led to her being taken to the Essex Assize Court in 1640, accused of plotting against the Protestant King. Although she was acquitted, worse was to follow. St Osyth Priory was searched, almost continually, for weaponry and many of her tenant farmers refused to pay their rents. Sir John Lucas, a popular and staunch Protestant royalist, was attacked at Colchester circa 1640 and St Osyth Priory was attacked by the local people. In January 1642, the Countess's sister spoke of being threatened daily by 'the people'.[115] In August 1642 the puritans rioted against Catholicism and Elizabeth became the most famous victim. Edward Hyde, Earl of Clarendon, included an account of her fate in his 'History of the Rebellion'.[116] The crowds completely ransacked St Osyth Priory. The Lady Rivers had already fled to Melford but the crowds followed her there and the Hall met the same fate as the Priory. In 1649 the Essex minister Ralph Josselin, on a trip to Melford, noted, 'I saw the ruines of that great (house), plundered out and desoloate (sic) without inhabitants', while St Osyth was later said to have remained uninhabited for the next seventy years.[117]

Countess Rivers found refuge with Catholic supporters in Suffolk and although she did not regain the trust of the people of St Osyth and Long Melford, Parliament ordered that her estates be restored to her. However, she came in to much debt and moved to France in May 1643. John Walter noted that the Countess put her losses at over £100,000.[118] About this time, a large portion of her Essex estates were sold, among them the Manors of Weeley, Great and Little Oakley, Skyghall, Beaumont, Holland, Fingringhoe, Mersea Island, Pete Hall, Kirby Hall and Thorpe Hall. On her death in 1651, she was completely bankrupt and her son John was left with nothing but a series of estates that were running at a huge financial loss. She was buried at St Osyth Priory.[119]

From the 1650s the lease of the Walton Hall estate was held by the Honeywood family of Kirby Hall. They signed the terms of the lease in 1652 following the death of Countess Rivers who had, as explained above, amassed large debts. These debts were the reason the manor of Kirby Hall was sold to the Honeywood family in the 1640s making it the first manor to be alienated from the rest of the Sokens. The Honeywoods originated from Marks Hall, near Coggeshall, and Kirby Hall descended with their family up until its sale in the early twentieth century. They leased the lands of the manor and were entirely non resident at Kirby, with the exception of one C.J. Honeywood, who lived at Kirby Hall between 1906 and 1908.[120] It was the Honeywood family who built the current red brick manor house, situated immediately behind the church and vicarage, circa 1720.[121]

Above: Kirby Hall. Image sourced from an original postcard. Courtesy of Richard Oxborrow, editor of www.kirby-le-soken.co.uk

Sir John Savage's time as lord of the manor of the Sokens was fairly abrupt and his involvement in Essex was minimal. Although his mother had bankrupted the Essex estates, the Savage family owned more land in Ireland and in the North West of England and it was there that he concentrated his efforts. Sir John was lord of the manor of Frodsham in the county of Cheshire as well as the Steward of Halton, also in Cheshire, which is where he lived, at Halton Castle. By basing himself in the North West he disentangled himself from the financial problems of the Essex estates, hoping that time would do its bearing and all the issues of past decades would be forgotten. Sir John still liked to be styled with his full set of titles however and he was known as Sir John Savage, 2nd Earl Rivers, 2nd Viscount Colchester, 2nd Viscount Savage and Lord

Darcy of Chiche. His other residence was the manor house at Rocksavage, Clifton, near Runcorn, Cheshire.

At the beginning of the English Civil War (1642-1651), in 1642, the 2nd Earl Rivers, who had reverted from a Parliamentarian to a Cavalier, was given a commission by the King to raise a standing army in Cheshire, which he did, and the Earl Rivers Regiment of Foote was born. It was large and well equipped with experienced officers and they were very successful in the north of England. John appointed his home at Halton Castle a fortified stronghold for the King and Captain Walter Primrose was put in charge. The castle was laid to siege twice over the period, firstly in 1643 when the Parliamentarians actually held the castle for a period, before abandoning it soon after. However, when the Royalists did not leave, the Parliamentarians surrounded the castle again and defeated it. Earl Rivers' other Cheshire residence at Rocksavage was left in ruins. Afterwards, he returned to Frodsham Castle, taking no other part in the war. He died on 10th October 1654, without even having visited his Essex estates. On the night he died, Frodsham Castle was set alight and it was with great difficulty that his body was rescued and later buried in Macclesfield.[122]

By 1654, the people of the Sokens faced an area that lacked prosperity, with much of the wealth passing out of the area and little investment taking place. Although international trade continued during the English Civil War, with luxury goods being imported and consumed and the elites not appearing to struggle financially there was certainly a degree of social and economic dislocation recognised by farmers and traders. The third Earl Rivers, Sir Thomas Savage, was imprisoned during the period before the Restoration on the charge of plotting to kill the Lord Protector, Oliver Cromwell. He was however freed in 1660, only to find his Cheshire residences in complete ruins, his father dead and the family estates in terrible attire financially. He decided to restore Rocksavage Hall as his family seat and lived not in the Essex estates but in London during the period of restoration. Walton Hall was leased on a ten year basis to Sir John Langham, a Baronet, in 1669 and following his death, to the Honeywood family of Kirby Hall in 1676 for £400.[123] Sir Thomas regained control of the estate in 1679. The house was then in the tenancy occupation of Richard Hewitt.

It is likely that is was under Sir Thomas that Walton Hall marshes were drained by a team of Dutch engineers under the supervision of Nicholas van Cropenbourgh.[124] The main reason for draining the marshes is likely to have been to create an increased area of secure sheep pasture, although draining it would also create the opportunity for change of use to arable cropping. The Naze had been a place of sheep grazing for centuries beforehand but such low lying marsh was forever at risk of flooding. Although the height of the English wool trade was in the fifteenth and sixteenth centuries, exemplified by the building of the famous 'wool churches' such as Holy Trinity at Long Melford, the Church of St Peter and St Paul at Lavenham and St Agnes's at Cawston in Norfolk, by the end of the seventeenth century wool still comprised two thirds of the total value of English exports and so the seventeenth century itself was a high period for sheep farming across East Anglia. Many sheep could be tended by relatively few shepherds and would be guaranteed to offer a valuable crop of wool every year, not to mention the money that came in from mutton and skins from 'excess animals'.[125] Prices didn't rise in the sixteenth and seventeenth centuries but prospects remained good.

The fourth Earl, Richard, or 'Tyburn Dick', as he was known because of a rumour that he tried to steal from his father when

he was a child, was next to inherit the Soken estates, in 1695.
[126] Sir Richard entered Parliament, under the title of Lord
Colchester, as MP for Wigan in 1681 and gained a commission
in the Horse Guards under Sarsfield in 1686. Like his father,
he was a staunch protestant and rode with William of Orange
on his way from Exeter to London in 1688. He was held in high
esteem at William's court.[127] Placing himself at the forefront of
the Williamite forces bode well for him and he quickly gained
promotion, becoming Major-General in 1698 and Lieutenant-
general in 1702. This followed successes for his forces in both
Ireland and the Netherlands. The 4th Earl was a military man
who lived for the action of the battlefield. He was
recommended by the Duke of Marlborough to command an
invasion force to France. However, this was eventually
diverted to Portugal and Rivers returned to England. Despite
this, he remained involved in military affairs and was granted
procurement with the cavalry. He also made sure to attend to
his political duties, especially as it became evident that
support for the Whig party was waning. He was appointed
Constable of the Tower in 1710 and was a popular figure at the
court of Queen Anne, appointed Master-General of the
Ordnance in 1711.[128] He married Penelope Downes in 1679,
daughter of Roger Downes, although she failed to bear him a
legitimate heir. He is however renowned for the amount of
mistresses he kept and he fathered several illegitimate
children, two of whom were with Anne, Countess of
Macclesfield. Rivers' intrigue with Lady Macclesfield was the
cause of her own divorce from her husband by Act of
Parliament in 1701. The poet Richard Savage claimed to be
the son of the 4th Earl and Countess Macclesfield but the
evidence in its support is not thought to be legitimate.

There is evidence that Richard visited Walton and stayed at
the Hall, as can be viewed in the memoirs of John Crosier
(Junior) of Maldon. Crosier noted that he spent two days with

Richard Savage at Walton Hall 'at the time of the fair', which suggests the 2nd July.[129] Crosier's memoirs provide an insight in to the society and industry of contemporary Walton. I quote below:

> From there we went to Walton in the Neys, where we spent two days agreeably at Mr Richard Savage's. It happen'd at the time of the Fair, when there is two nights dance. All the country lads and lasses meet, with whom we join'd. I cannot say much as to the beauty of Walton, but its society and neighbourhood compensates for every other deficiency. The Coperas House we took particular notice of, and indeed it's very curious. It's a peculiar kind of earth which is found and gathered on the sea shore; then laid on a large space of ground under which are deep spaces for the reception of the liquor which drains from this earth. Then, it's pumped into coppers and boil'd; then let into cisterns where, as it grows cool, it candy's on sticks and is fit for use. It's of a beautiful transparent blue. We left our Walton friends with some reluctance...[130]

In June 1712 Rivers was promoted to the rank of General, and became commander-in-chief in England; he died a few weeks later, at Ealing Grove on 18th August 1712.[131] As Rivers did not have a legitimate son, the Earldom, and the lordship temporarily, passed to his cousin John Savage, a priest in the Roman Catholic Church, on whose death, circa 1735, all the family titles became extinct. The estates were then held by the trustees and executors of the 4th and 5th Earl's estates, being Robert Harley, 1st Earl of Gloucester (d. 1724) and Charles Talbot, Duke and 12th Earl of Shrewsbury (d.1718).[132]

The manors eventually passed to Richard's illegitimate daughter Elizabeth, or 'Bessy' as she was known. An Act of Parliament had to be passed to confirm her rightful succession and this came in 1721. However, Bessy had married the Rt Hon. Frederick Nassau de Zuylestein, 3rd Earl of Rochford, in August 1714. He took control of the lordship. When he died in 1738 Bessy later married the Reverend Phillip Carter, again granting the lordship to him. Upon her death in 1746, the estates descended in the Rochford line and were taken by her son, William Henry Nassau de Zuylestein, 4th Earl of Rochford.[133] William was born at St Osyth Priory in September 1717 where he spent his childhood although he was educated at Eton College from 1725-32.[134] In 1746 he became the lord of the manor of many Essex estates.[16] However, due to the nature of his work, he was non-resident in the Sokens for much of his life, perhaps leading to his decision to dispose of particular estates, including Walton Hall, in 1775.

The 4th Earl spent his life engaged in diplomatic work abroad. However he was also involved at court where he was a Lord of the Bedchamber from 1738 and Vice-Admiral of Essex from 1748. In 1749 he was made envoy to Turin and remained there, a keen linguist, until 1755 when he became a Privy Councillor and a Lord Justice of the Realm. He served as lord Lieutenant of Essex from 1756-1781.[135] He served as ambassador to Spain at Madrid from 1763-66 and to France at Paris from 1766-68. He returned to English domestic politics after leaving France, becoming Secretary of State for the

---

[16] Property held by the 4th Earl of Rochford in 1746: 'The manors of Chick St Osyth, Wiverhall or Wickborne in Chick, Great Clackton, Little Clackton, Kirby, Thorpe, Walton and Farshall. He also held 24 messuages, one mill, 30 gardens, 3,800 acres of land, 530 of pasture, 1430 of marsh, 850 of woodland, 200 of furze and heath, 1/6 of the manor of manor of Can Hall with the apparts...and the Advowsons of the Churches of Great Clackton, Little Clackton, Little Holland, Thorpe, Kirby and Walton. He also held extensive overseas estates, especially in the Netherlands.'

North 1766-70 and the South 1770-75. It is often noted that he had the casting vote in the decision bill of whether to withdraw the 'obnoxious' tax system in the American colonies in 1775. He voted against the motion, which led directly to the American War of Independence.

Despite having to sell the entirety of the Soken estates in 1775 to alleviate his debts, in that year Rochford was appointed Master of Trinity House which he remained until 1777. In that role he granted the new landowner of the Naze, Richard Rigby, permission to use the Trinity Tower for 'his own purposes'. The 4th Earl was made a Knight of the Garter in 1778 but died childless and was succeeded in his peerage titles by his nephew, William Henry, the 5th Earl. It was due to the accumulation of significant debts that William was forced to sell the manorial rights and lands of the Sokens in 1775. The purchaser was the Honourable Richard Rigby of Mistley Hall.

Rigby was an aristocratic MP who lived a flamboyant lifestyle. He was Member for Castle Rising, Norfolk in 1745, Sudbury in 1747 and Tavistock in Devon from 1754-84, although he rarely visited those he represented. Rigby was born in to a wealthy family, established at Mistley Hall, near Manningtree, Essex.[136] The family had prospered from the colonial cloth trade and had also done well speculatively in the South Sea Bubble. Rigby's father had been Secretary of Jamaica, a substantial landowner in Antigua and a slave trader. His grandfather, Edward Rigby, had inherited the village of Mistley in 1703 from Aubrey de Vere, the 20th Earl of Oxford. Richard Rigby himself was educated at Corpus Christi College, Cambridge and the Middle Temple before following in to the life of politics that surrounded him throughout his life. He was committed to his local area however and invested heavily in it. The modern day village of Mistley was almost entirely built up by Rigby, including many of the shops, the dock and warehouses. Rigby had a vision to develop Mistley into a Spa town on par with Bath and Harrogate. He employed the architect Robert Adam to design the venture which was to prove to be his downfall. At the same time Lancelot 'Capability' Brown was employed to design the gardens at Mistley Hall. Work began in Mistley in 1776 and carried on until Rigby's death, although the project was never finished.

Rigby's first role as MP led him to represent the constituency of Tavistock in Devon, whose electorate at that time was around 30 people. The normal means of becoming an MP in the eighteenth century was to pay the 'owner' of the Parliamentary seat, in this case the Duke of Bedford, for the privilege of representing it. In the case of Rigby, he did not have to pay due to the numerous favours he had done for Bedford. Rigby was himself a 'Bedfordite'; a member of the Bedford Whig political faction that existed in Parliament in the third quarter of the eighteenth century under the leadership of John Russell, 4th Duke of Bedford.

Rigby spent most of his time in his parliamentary office at Westminster, although he regularly entertained guests at Mistley Hall and at the Naze Tower. He was an 'empire builder' and his ambition was to travel from Mistley to Walton Hall without leaving his own estates.[137] He achieved this ambition in the late 1770s. Rigby was the first proprietor for many generations to have spent periods of time living at his Walton Hall estate. He enjoyed the individual character of the Tower and invited Georgian aristocrats to visit him there, hosting particularly raucous 'tea parties'. He invested large amounts of capital in the estate and expanded the house. In 1778 his estate in the Sokens amounted to around 1900 acres including 455 acres at Walton Hall, 774 acres of saltings and 434 acres at Thorpe Park.[138] Walton Hall then consisted of the entire Naze headland and much of the land where the present town lies, along with the Hall and Trinity House Tower. John Barnard leased and farmed the 455 acres of usable land on the estate for the entirety of the Rigby premiership.

In June 1768 Rigby was appointed to one of the most lucrative governmental roles, that of Paymaster General to George III, a cabinet-level post with responsibility for army pay, rations and logistics. It was a post considered to be one of the most important in the government of the time and was, as suggested, open to corruption with the handling of large sums of money and the ability to charge interest on loans disbursed. This ability to bend the rules led to significant problems for Rigby. In 1776 he faced the issue of both large numbers of soldiers stationed in the American colonies after they declared independence in 1776 and the fact that several cabinet members were starting to look twice at the budget of the Paymaster General and wondering where the money was heading. It was found that many of the contracts for machinery and warship building had been granted to the docks and factories in Mistley, the profits of which went directly to Rigby himself. When the British forces were defeated in 1781, a review was undertaken of Rigby's department which led to his resignation shortly afterward. His pay was frozen and funding for the project in Mistley ended. His reputation amongst the political class was destroyed. In the mid nineteenth century Conservative Prime Minister Benjamin Disraeli, in *Coningsby* (1844), gave the name Rigby to a 'political intriguer and parasite'.[139]

On 23rd February 1784 Rigby was required to pay interest on outstanding balances from his tenure as Paymaster: four years later he still owed £156,000.[140] The review had assumed that he had been taking advantage of the system in place, whereby one only had to ask the Treasury for a certain amount of funds and they would be granted. Rigby had been sending the capital to Mistley to fund his personal projects. He was disgraced and died shortly afterwards on 8th April 1788, leaving his estate, numerous angry creditors and a host of other problems to his sisters Martha Hale (d. by 1799) and Anne Rigby (d. by 1799), and his nephew Colonel Francis Hale who adopted the name Francis Hale-Rigby. When his will was proved in 1793 the full extent of his debts to the Crown became apparent. His executors, Timothy Caswell, Daniel Macnamara and his nephew Francis were left with the duty of dealing with his debts to the crown being £151,783 3s 6d, with his remaining assets amounting to just £40,408 11s 2d.[141]

To make matters worse for the family situation, a lot of his existing wealth was gifted to his daughter who had married General Hale and ultimately it went to the Pitt Rivers family. Seeing the extent of the Rigby debts, Francis took the decision to sell off the Soken manors to various smaller landowners. After fifty years the Mistley estate itself was broken up and auctioned. The two towers on the riverfront, which are all that remain of Rigby's florid Church, and the Swan fountain further to the east, are all that remain of his grand spa town scheme. From 1778 the Walton Hall estate had been let to Mr John Barnard who remained there until 1791 when William Salmon became tenant, leaving in 1798. The Soken manorial rights were sold by Francis Hale-Rigby to Edward Russell Howe in 1801. The agricultural lands and saltings of Walton Hall were sold separately thus alienating the manor from the lordship.[142] The indentures of the execution of Richard Rigby's estate are located at the Essex Record Office and provide evidence of what was quite a common problem regarding the inheritance of large scale debt through legacies. Sadly, Richard Rigby does not seem to have realised the severe problems he was handing to his heirs and in his last will and testament, dated 31st December 1781, he makes no reference to his severe debts but still grants substantial amounts of money to acquaintances.[143]

The manorial rights of the Sokens originated when St Paul's had full 'liberty' of the Sokens and meant that the lord of the manor had the greatest freedom to exercise his exact wishes on the Naze. Although individual lords of the manor had little direct influence on the way people lived, they played an instrumental role in the maintenance of the local legal system.[144] In the Sokens, the lord's jurisdiction extended over the two courts that existed within the soke until well in to the nineteenth century: the court baron and the ecclesiastical court. The court baron was mainly concerned with land transactions that took place within the manor but also property offences and sometimes more serious offences regarding breach of law. It met once a year at the minimum on St Anne's day (26th July), usually at Kirby Hall but later often in the church in Thorpe.[145] The ecclesiastical court was more concerned with church affairs including wills and marriage certification. According to Morant this met *'every three weeks, as occasion serves'*.[146] The final court session of the Soken court was held on 2nd July 1861 at Thorpe church following its order of abolition.[147]

The manorial rights were sold relatively frequently in the nineteenth century, at a time when such denominations counted for an increasingly little amount. Charles Pearson paid £4600 for the rights in 1804.[148] Pearson was a forward looking entrepreneur who, it is thought, purchased the lordship to obtain rights of collection of copperas and materials for making cement from Walton's foreshore. These industries became very lucrative for Pearson and he reserved the rights of collection when selling the manor to the Revd Thomas Scott, rector of Little Oakley and Wix, in 1811.[149]

The collection of copperas was linked to an unusual industry closely connected to cloth manufacture. Copperas can be used to create a black dye, which had a particularly high demand in Portugal, the principal export destination. The copperas, in the form of iron bisulphate, was washed from the cliffs by the waves and was gathered from the foreshore on the eastern side of the Naze by many women and children. Copperas was mixed with scrap iron, and vitriol and sulphuric acid were produced, both highly dangerous substances. There were originally two copperas works in Walton, one destroyed in 1725, the other, in Vicarage Lane, closing in 1855. However, copperas continued to be pillaged through the 19th century and copperas stones were collected from Walton's foreshore until at least Christmas 1909.[150] Stan Jarvis has written that the ground at the Naze became significantly polluted with sulphur, so much so that a hundred years after extraction was halted nothing grew on the cliff face.[151] Although a profitable industry, it was eventually not so for Pearson who was declared bankrupt in 1835.[152]

The other industry connected with the Naze cliffs was the collection of septaria for 'Roman cement'. Septaria is an argillaceous limestone which occurs in the London clay and can be used either as a natural building stone (the Romans made great use of it) or as cement. It was James Parker of Northfleet who began the process of making 'Roman cement' as it was called, in 1796.[153] It was made by burning and crushing the septaria and was much stronger than other cements, even setting under water. Over two million bushels of septaria based cement were made at Harwich in the dedicated factory there over the course of its manufacturing lifetime, yielding 30 to 40 shillings a tonne.[154] At the height of the Walton septaria extraction industry at the Naze, James Pigot noted in his Directory of 1839 that 'the shore at Walton-le-Soken abounds with pyrites and nodules of argillaceous clay, which, when they have acquired firmness, are collected and conveyed to London and Harwich for making Roman cement.'[155] The great extent of the digging for septaria meant that erosion became an increasing problem leading to siltation among other coastal issues. In 1845 the Commission on Harbours of Refuge recommended that removal of septaria stone be stopped which was done. The national industry began to steadily decline in the 1870s as Kentish Portland Cement surpassed cement stone. Portland cement was readily and easily available in the chalk quarries of Rosherville, Swanscombe and Northfleet.[156]

The Revd Scott sold the manorial title in 1826 to Benjamin Chapman, a solicitor who lived in Harwich.[157] He died circa 1835 and was succeeded by the trustees for his widow Fanny Chapman, being his son Edward Chapman, and Richard Read-Barnes, who was proclaimed steward of the estate in 1826 and was Borough Treasurer of Harwich. Fanny Chapman died some time before 1851 and Barnes in 1852, leaving Edward Chapman of Harwich, a Walton Improvement Commissioner, as sole trustee and lord of the manor.[158] He died in 1893 and was succeeded in 1894 by Fanny Fisher Chapman, Charlotte Chapman, Richard Saxty Barnes, and George William Jones of Fannington, Suffolk, trustees under his will. Fanny Chapman had died by 1897 and Charlotte Chapman by the following year.[159] In July 1900, a local dignitary, Colonel Richard Percival Davis, of New House Farm in Walton, showed an interest in purchasing the titles which would have brought upon their return to Walton. Davis was High Sherriff of Essex, a Justice of the Peace and a County Councillor. However, according to his solicitors, Woodroofe and Burgess of Lincoln's Inn, he turned down the offer on account of the asking price of £2,500 being, in his eyes, a 'ridiculous sum to pay'.[160]

Barnes and Jones eventually sold the manorial rights for the same amount, £2500, in 1901, to Henry E. Paine and Richard Brettall of Chertsey, Surrey, both solicitors. Beaumont and Son of Coggeshall were appointed as stewards and administrators of the Soken manors. Paine remained lord until circa 1918.[161] By 1921 the manor had passed to Emma Elizabeth Freeman of Chertsey, Surrey and Clara Freeman of Canterbury, Kent, who still held the title circa 1927. Descent of the lordship is then not documented until 1934-5 when it was held by Clara Freeman and Owen Warner. Guy's Hospital is thought to have owned the title at points during the twentieth century although it was sold to L. Maddock of Mettingham Castle, Suffolk, in about 1963.[162] The title has come up for sale a number of times since the 1960s.

Unlike the lands of Kirby Hall and Thorpe Hall, which were disposed of in the mid seventeenth century, the Walton Hall estate was only divorced from the manor on the sale of Richard Rigby's estates in 1801-2. Other than between 1791 and 1798 when William Salmon was tenant, John Barnard of New House Farm was tenant of the Hall estate between 1778 and 1801 having paid an annual rent of £350.[163] His son Benjamin purchased the freehold for £18,000 (about £750,000 today) on 18th August 1802, at the beginning of the sale period of Richard Rigby's extensive Tendring estates, by his heirs Martha Hale, Anne Rigby and Francis Hale-Rigby.[164/17] In the

---

[17] Walton Hall is noted as having contained, in 1801, 326 acres of arable land, 60 acres of marsh, 750 acres of salting and 45 acres of woodland. The original sale particulars of 1801 describe 'the tower' (presumably the tower that is now connected to the house as opposed to 'the Naze Tower') as having been 'fitted out by the late proprietor at considerable expense - £1500 – and made a complete shooting seat with kitchen, cellars and servants rooms, fit for a small genteel

case of Walton Hall, the saltings were thought to be of no monetary worth and Barnard therefore took them as part of the sale.[165]

The system of copyhold tenure was abolished in the mid nineteenth century and replaced universally with a freehold system of ownership. This early divide between lordship and lands following Rigby's death was therefore visionary at the time, even if brought forth by Rigby's executors through necessity rather than choice. It appears from the documents that Benjamin Barnard ultimately leased the estate and then rented it from the leaseholders so as to regain some of the value. There are lease documents showing him signing it to Francis Hale Rigby on 24th December 1802 and to Wakelin Welch of Lympstone, Devon on 27th September 1806.[166] This action was likely to be due to Barnard wanting to regain some of the financial value whilst he continued to work the land with his family.

Although there is no documentation to prove it, it is likely that it was Barnard who expanded the barn complex, a project furthered by his son William and probably also the Hicks family. He was determined as to how he wished his estate to be run following his death, laid out in a precise manner in his last will and testament of May 1813. He placed great emphasis on the keeping of good accounts; something it appears that was not always done. I quote directly below:[18]

> I give the farm of Walton Hall to William Barnard for the benefit of himself and all my other children. William shall carry on and manage the said farming business, shall be employed for that purpose and it is

---

family, both for a summer residence and a sporting seat in Winter.' Barnard bought the estate at auction in 1802.
[18] The will can be found at the Essex Records Office.

my will further that proper books of account shall be provided and kept in some convenient part of my present dwelling house (Walton Hall house) wherein shall be entered an account of all monies paid, disbursed and received on account of the said farming business for my son William and to which all my other executors shall have free access to receipts to be made after each and every year following my decease until the expiration of the current lease. *167*

William was granted a salary of £200 a year.[168] The executors of the will were William and Benjamin (junior) Barnard, sons of Benjamin as well as Richard Stone of Frinton Hall and William Dennis of Kirby-le-Soken (who married Benjamin Barnard's daughter Ann on 12th May 1814).[169] Benjamin died on October 19th 1813.[170] William had married Jemima née Sallows in 1811 and he continued to farm the estate for the remainder of the seven year lease after Benjamin's death. Following this particular lease period Benjamin Barnard had ordered that the estate be sold with capital to be shared between siblings.[171] The sale was done and a new owner found, Mrs Elizabeth Welch, widow of Mr Wakelin Welch who had previously leased the estate. William Barnard died in 1822. The executors of his will, particularly Richard Stone of Frinton Hall, took responsibility for the management of the estate until the new owner could organise an alternative system.[172] However, Stone discovered that William Barnard had been mismanaging the accounts. [173] The early nineteenth century was far from an easy time for farmers with wheat prices fluctuating substantially. In 1812 they were as high as 126 shillings and 6d a quarter but in 1816 they had fallen to just 78 shillings and 6d. By the late 1860s prices had dropped to between 40 and 56 shillings per year. Such differences were the result of political turmoil particularly due to the repeal of the Corn Laws in 1846, as well as war and the weather which all influenced prices and thus the livelihoods of those who relied on the land for their income in an age before subsidies.

A convenient marriage between John Woodgate Hicks and Rosa Barnard in 1817, ultimately led to another significant farming family of the area, the Hicks, becoming involved at Walton Hall. The Hicks family farmed at Great Holland as well as in Kirby, Frinton and Thorpe during the Rigby period at Walton. J.W. Hicks' father, John Timperley Hicks, had been tenant at Great Holland Hall from 1800 until 1829 when his son Charles took over the farming operations there. Charles had previously farmed at Dairy House farm in Great Holland.

Above: Great Holland Hall c.1870. Image sourced from an original postcard kindly provided by Linda O'Reilly.

Mainly due to his position in charge of sea defences on behalf of the Commissioners for Sewers of the Tendring Hundred Level Richard Stone was able to network with farmers across the district and therefore came to know John Woodgate Hicks well.[174] Stone regularly inspected the walls on the Naze in his capacity as local inspector, noting a survey on 23rd May 1816 resulting in a measurement of the 'new wall' as being four miles and forty three rods, twenty one feet.[175] On 18th March 1827 he wrote in his diary that the 'tide did considerable damage to the old wall (at Walton Hall), about forty rods nearly through beside great damage in Broad Piece'.[176] Further, on 31st October 1827 he stated that a 'high tide in the night did much damage at Walton Hall' and he was in charge of dealing with the resultant problems. On 18th February 1828 he went to Walton Hall to 'set out wall horses in broad fence' and on 21st August 1829 to 'look at the wall at Walton Hall and put a new horse at the corner of the old wall 42 piles to ask 138'.[177]

Stone knew that Hicks was interested in taking on Walton Hall but it was the burden of sea wall maintenance that concerned Hicks most before he could make the decision of whether or not to lease the Naze estate. On 21st March 1827 Hicks stayed at Walton Hall along with his father, Richard Stone and Elizabeth Welch. Stone indicated to Elizabeth Welch that Mr Hicks was 'determined to have nothing to do with the Hall unless Mrs Welch keep up all walls and guarantee all damage from inundation.'[178] It was a big responsibility but it is clear that Welch wanted a reliable tenant and Hicks would be just that. After a final meeting with Stone at Colchester on 15th September 1827, Hicks signed the lease of Walton Hall at Sproughton, for '13 years at £750 per year with Mrs Welch to do all the walls and keep them in repair JWH doing twenty days carting with a team and a horse and cart occasingly (sic). Mrs Welch to put the premises in repair, after which Mrs Welch to perform usual covenants'.[179]

More investment also went in to expanding the barn complex. The Hicks family lived at Wormingford until 1828 when reports of new births at Walton Hall seem to suggest they had moved there permanently.[180] Stone wrote in his diary that 'Mr and Mrs Hicks came to reside at Walton Hall' on 2nd October 1827.[181] It seems that the family continued their lease of the Hall for another 50 years. On 5th September 1848 the estate was purchased at auction by one Richard Stone or 'Stong (sic)' for £17,000.[182] This may have been a relation of the diarist and farmer Richard Stone. The Hicks family continued to farm the land until 1864 when Mr Peter Blofield succeeded them.[183] In 1863 the estate belonged to a Miss Mary Powell, although it is not known what her relationship with the former owner was, or how she was connected with Walton.[184]

Many of the barns, including the old granary shown above, were constructed during the Barnard and Hicks eras of management. Above: Photograph courtesy of Putmans Photographers - The Walton Archive – www.putmans.co.uk

In that same year, 1863, Sir John Henry Johnson J.P, purchased St Osyth Priory at auction, as well as most of Rigby's former Soken estates. The Priory had been uninhabited since the Nassau family departed in the late eighteenth century and had fallen in to disrepair. Sir John began a new grand building scheme there, demolishing some of the buildings and replacing them with rich and lavish entertaining rooms as well as converting the monks' dorter into a chapel. He also undertook a grand plan in the gardens and masterminded several of the designs himself. Johnson purchased the Walton Hall estate from Mary Powell at auction in November 1872 paying £24,000.[185] The farm and house was let to John Woodruffe Eagle, who had recently moved from Wix Abbey.[19] Rights to remove materials from the foreshore remained one of the privileges of the lord of the manor. Most late 19th century lords ignored this and so farmers could use materials freely for sea defence. In 1906 J.W. Eagle explained that he had, for 33 years previously, freely taken shingle from the foreshore for road maintenance and sea defence but was now being asked to pay £10 for 200 loads.[186]

---

[19] The auction was held on Thursday 11th July 1872 at the Mart, Tokenhouse Yard, London. The sale catalogue notes the holdings to comprise 'a substantially-erected residence, occupying a commanding site, on high ground, a suitable homestead, four cottages for labourers (Creek and Walton Hall cottages were not built by this time and the cottages were situated near to High Tree lane), surrounded by enclosures of arable land of unusual fertility, and fine grazing and meadow land. The whole embracing an area of 1,188a 3r 18p.' There is also noted as having been 'an abundant supply of fine spring water', a supply that we must assume has been diverted from collection. The Hall then comprised 'entrance hall, dining and drawing rooms, study, nine bedrooms, nurseries, excellent domestic offices, brew house and two supplementary chambers. There were also gardens, nag horse stabling, carriage house, an extensive range of barns with lean-to boxes and stables, several enclosed cattle, horse and sheep yards with open sheds, granary, wagon, cart and implement sheds, stables for upward of twenty cart horses, with harness and chaff houses, loose boxes, cow and calf houses, stock yard, home paddocks and four capital built cottages.'

Sir John financed the establishment of a coastal defence system at the Naze with breakwaters to act as sand retainers and cliff drains to minimise land slips, the remnants of which can be seen on the beaches or protruding from the cliffs.[187] By doing so he added his name to the many before and after him who have tried to mitigate the effects of erosion at the Naze.

Arguably the most significant project to never actually happen at the Naze was the plan to connect Walton with Harwich by rail, bridging the eastern end of Hamford Water. This idea rose in 1881, part of Phillip Brannon's failed attempt to develop the Naze end of Walton. A large map entitled 'Walton and Harwich Junction Railway in Parliamentary Session 1881/1882' was recently discovered by local historians at the Council House in Old Road, Frinton which was the beginning of an enlightening discovery regarding the extent of the plans which would have encircled the Sokens with railways. The map was produced by William Dennis and Alex Thuey (the engineers) and Phillip Brannon was the architect. The plan entailed the railway to be built north of Walton station and the Parish Church, at the junction of Kirby Road and Walton High Street, before going up to Warner's Dock and up the west coast of the Naze looking out across the backwaters. It crossed Cormorant Creek and the saltmarshes before leaving the mainland at Stone Point and crossing to Great Oakley, just south of Foulton Hall.[188] This would have been an incredibly ambitious engineering project although several contemporaries appeared to show support. Peter Bruff, who was central to the development of Walton as a Victorian coastal resort, was involved in the project proposals. He wrote a report on the 'Felixstowe Dock, Harwich and Walton Railway' in January 1882, presented to the Harwich Harbour Conservancy Board in London. The Board received the report very positively and the Chairman, Colonel Tomline, is reported as saying 'we have no objection to it'.[189] The ambition of the project at a time when railway finances were poor seems to be the main reason why the project was ultimately abandoned. However, local opposition due to fishing rights was also a consideration as was the failing health of the key drivers of the project – sadly all conjectural reasons as any sources regarding the project appear to disappear post 1882.

After Sir John Johnson's death, the Naze estate was held by trustees to his will from circa 1907 to 1920. J.W. Eagle died in 1916 but his son, another J.W. Eagle, continued the family business and took the decision to buy the estate. He had occupied the house from circa 1908 when a new lease was signed on 6th February. [190] On 10th December 1920 he purchased the Walton Hall estate as well as Hedge End Island from the trustees of Sir John Johnson for £8500.[191]

Above: J.W. Eagle (1842-1916) in the farm office at Walton Hall. Courtesy of the Eagle family.

John Eagle *(1885-1969)* was a significant landowner in the Tendring Hundred district. Through the 1920s, in addition to Walton Hall, he bought the Kirby farms of Nortons Farm, Marsh Farm, Brick Barn and Devereux Farm as well as the Ashes Farm, on the boundary between Kirby and Walton, acquired from Colonel J.P. Davis. Eagle was heavily involved in the local community in Walton. He assisted in the efforts of the local lifeboat crews, rising to the position of chairman of the local branch. He was also a Justice of the Peace as his father was before him, Chairman of the Walton Urban District Council, the local mayor for a period and was prominent in the Tendring Hundred farming community. From 1920-24 he was Commodore at the Walton and Frinton Yacht Club. He married Dorothy Franklin on the 14th June 1916 at St James' Parish Church, West Hampstead, London and they had one child – John Woodruffe (1922-2005), who eventually took over as the principal partner of the farm business. As a man John Woodruffe senior was sometimes difficult to get along with. In the words of his daughter in law, my grandmother, Daphne Eagle:

> Woodruffe was a real Victorian. I remember I brought Dorothy back from having an operation in hospital and Woodruffe was sitting in a chair. He said "Oh good. Dot, get me a gin and tonic". I exploded and told him to get his own gin and tonic! But, that was him. He didn't know any better...He lived for the Council, the golf club and poker. He was a J.P and Chairman of the Council. He was Woodruffe.[192]

In November 1924 Woodruffe sold a large part of the eastern part of the Naze to Mr Ernest Alexander, including Hall lane and Naze Park Road. In February 1928 Mr Alexander opened a Golf Club, situated where the public open space and Nature Reserve are today. It was designed by James Braid, a prominent golfer who had been British Open Champion five times between 1906 and 1914. Braid was one of the foremost recognised golf course architects and designed dozens of courses across the country. His vision for the Naze involved an eighteen hole layout of 6250 yards with seventy two bunkers and eighteen holes of varying degrees of difficulty. The course proved very popular, rivalling the course at Frinton, and boasted of having 'a complete view of the sea from all areas of the course'. A long, single story clubhouse was built, thatched with Norfolk reeds and permission acquired from Trinity House to make use of the ground floor of the Naze Tower as a 'shelter only'.[193]

The Club grew rapidly from the outset and within two years had in excess of 800 members. At its peak in the period 1935-1938 it had over 1200 members including Lord Byng, Sir Arthur Christiensen, editor of the Daily Express, Sir Albert and Lady Clavering and Kathleen Garnham who played internationally. Lord Tennyson and Edward Prince of Wales (later Edward VIII) played at the Club on at least one occasion. However, following a period of drastic economic depression for the Club and the wider country in the 1930s, Ernest Alexander was declared bankrupt in 1938 and the Golf Club was closed. The land reverted to agricultural use, being ploughed and sown with rapeseed, before it was requisitioned by the armed forces in 1939. The beaches were mined and barbed wire and anti-tank defences were erected. The clubhouse was used as billets during the war but was sadly struck by lightning and burnt to the ground in 1946.[194] This was probably one of the key reasons why the Club was never resurrected although other reasons included the neglected sea defences which were regularly breached after the war and resulted in many areas of the former course being covered at areas of high tide. In 1953 the 'great flood' completely destroyed a number of holes that were still apparent and removed any possibility of the Club beginning again, setting the scene for a number of planning applications which if they had been successful would have entirely changed the character of the landscape.

The Eagle family continued to farm at Walton Hall along with Walton Ashes and in Kirby-le-Soken both during and following the war. J.W. Eagle's brother, Frank Douglas Eagle also farmed land in St Osyth. The 'Naze Links Estate', consisting of the former golf course, today the public open space and nature reserve, was owned by A.H. Tucker from shortly after the Second World War to 1963 and he permitted the Eagles to graze the land in to the 1960s, free of charge. It was in the 1950s however that public concerns for the future of the Naze were first raised, perhaps because it was the first time that the site faced the prospect of significant change. For centuries the Naze had been agricultural, with the exception of the short period of time as a golf course and military encampment. In 1955 the local council gave permission for the construction of 'a promenade, a car park for approximately 200 cars, beach hut sites, three lavatories, a public shelter for residential buildings, refreshment room and kiosks, an area for residential building sites with service road and the provision of gardens, the whole comprising public recreation grounds and amenities, situated at the Walton Golf Links Estate, Walton-on-the-Naze'. However, certain conditions were placed on the permission, including a requirement to provide a great deal of more detailed information. The applicant was Tucker himself but development never actually began. Instead, the Naze faced a number of years of uncertainty. In 1958 an application by a prospective purchaser for using the Naze Links estate as a caravan site was refused. In 1959 there were two applications, the first including a proposal to build 67 houses, 445 chalets, large numbers of beach huts, a car park, lavatories, gardens, bathing pool, tennis courts and bowling green. The second application was entered by the 'Naze Development Company' which set out to extend the development of the original 1955 application to include a motel and a 'refreshment room/restaurant' as well as fifty detached houses, four lavatory blocks, a sea wall and greensward and to extend the car park to include up to 25 coaches and 490 cars. This similarly failed and the land continued as grazing pasture.

However, concern was sufficient for a public campaign to 'save the Naze' to build momentum. A 'Save the Naze' Association was created, based at the Council House in Old Road, in Frinton and a campaign triggered, involving the national press, to take the Naze in to public ownership. A petition was created including multiple signatures from across the country and letters sent to F.A Skidmore who led the campaign. One such letter reads as follows:

29th February 1960

F Skidmore, Esq.
Chairman
"Save the Naze Association"
Walton on the Naze

Dear Sir,

My wife and I have followed with interest the reports of your activities in the national press. We support your objects very fully, and would like to associate ourselves with your organisation. If there is anything we can do to give strength to your cause, please let us know.

Yours sincerely
Stewart Platt

Another letter read...

6th April 1960

Dear Sir,

I am now a Clacton resident and would like to add my name to those folk petitioning to SAVE-THE-NAZE. There are so few places as unspoilt. Walton's Naze has saved my life when I stayed there two years or so ago.

Yours Faithfully
Miss C.E Crookes BA (cantab)

In 1961, a consortium, led by the former England cricketer Godfrey Evans, approached the landowner with a plan to develop the 78 acres surrounding the Tower with 350 permanent dwellings, holiday beach huts and a motel.[195] Evans' grand plan would have changed the landscape most significantly but was rejected by the Frinton and Walton Urban District Council following a campaign by the public in April 1961. Evans appealed and a government inspector, Mr H.W Lovell, was sent from the Ministry of Housing and Local Government, to review the application. Lovell agreed with the council and suggested the Minister reject the appeal, which was done. The Inspector reported that the various past permissions that had been granted, along with Evans' application would 'change the character of the southern end of the site to such an extent that it could no longer be regarded as an open space, and such development would have a considerable impact on the remainder of the area.'[196] Lovell's name has since been forgotten by most but in terms of the Naze's landscape history he played a significant part, for it was his decision that safeguarded the Naze as the site it is today, 'unblemished' by built development. For Lovell it seemed that the Naze's situation, scientific interest and amenity value placed it in a special category. Any development of the Naze would destroy this character and, as land for residential development would apparently soon become available elsewhere, he saw no overwhelming reasons why development on the scale proposed should take place at the Naze.

During the early 1960s, when the struggle between Evans and those against the development proposals was occurring, a very different struggle was taking place on the Naze itself, between a man and the elements. George Smith lived at the Naze in a tumbledown shack for a number of years and according to Isobel Ruckert, was well respected by the Old Hall Lane/Sunny Point community at the Naze. Isobel's mother and the Matron from the Samuel Lewis Convalescent Home would give him food. However, sometimes he suffered from abuse. A cutting from the East Essex Gazette from 1962 notes that 'even Walton's hermit, George Smith, who has been living on the cliffs at the Naze since February, has not escaped the attention of hooligans. A group of teenagers, leaving their van at the top of the cliffs, threw stones and sticks at his Robinson Crusoe home, denting his precious one and only kettle. George is thinking of leaving the area to find somewhere with a safer degree of solitude.'

In 1963, the eastern most edge of the peninsula was bought by Essex County Council for £75,000 as a 'public open space' (Frinton and Walton Urban District Council paid 30% of the cost) and control was transferred fully to Frinton and Walton Urban District Council late in 1965. A sub-committee was established by Essex County Council soon after the purchase in 1963, to plan the future management of the Naze, most of which has not changed since their decisions were taken early in the 1960s. In 1965 they recommended the following:

1. That only 'minimum coast protection and cliff preservation works should be undertaken and that a promenade should not form part of any future scheme.
2. The permanent car park should accommodate 1000 cars.
3. 'That suitable tree planting be carried out in order to screen the 'refreshment and toilet blocks' from any future dwellings to be erected on Old Hall Lane.'
4. 'That an area of approximately 9.3 acres be designated as a Nature Reserve subject to details of the same and its control being agreed following consultation with interested societies.'
5. That certain 'rehabilitation works' take place including removing 'any unnatural rutting on various parts of the land but that the large mound of earth to the west of the Naze Tower be left, to be ultimately spread in the gardens of any residential dwellings to be erected on the site. Remove any derelict buildings, not including military buildings such as pill boxes. Concrete gun bases should be retained as public seat bases. Old breakwaters should be removed as a hazard. Any dangerous trees should be felled. Suitable notices should be erected.'
6. That a new grazing tenancy agreement be entered into with Messrs JW & FD Eagle on a yearly basis on approximately 17.3 acres of land at the southern end of field 5183 being bounded on the eastern side by the eastern boundary of the existing agreement and including part of field 3686. The farmer should also be instructed to remove all barbed wire fences and replace them with plain wire fences on the boundaries of the new area.'
7. The Tree Warden of the County Council should prepare a plan of a tree planting scheme for the whole of the Naze.

8. That the area be known as 'The Naze'.[197]

Soon afterwards Chairman of the Council T.G. Rainbird sent out an appeal to plant a thousand trees at the Naze, particularly targeted at 'lovers of the countryside' to allow them to be involved in 'improving the beauty of the area'. Suggested trees included common and grey alder, ash, silver birch, common broom, hornbeam, mountain ash, English oak, sea buckthorn, sycamore, spindle, may or silver/violet willow. In 1970 the council approached the Essex Naturalists' Trust (now Essex Wildlife Trust) to form what is now the John Weston Nature Reserve, dating officially to 1971.[198] John Weston was intimately involved in the protests against development of the site in the late 1950s and was the first warden of the reserve. In the words of one of his protégés Weston was 'one of the best all round naturalists that Essex has ever produced'.[199] He was an environmental activist far ahead of his time and dedicated much of his life to protecting, observing and recording wildlife around Hamford Water and the wider county. Weston organised a long standing Little Tern and shore nesting bird wardening scheme at the Naze which began in the early 1970s. About thirty five people took part each year with regular volunteers named through correspondence as Weston, Clarke, Brett, Hobern, Brooks, Davidson, Manning, Beach, Haggis, Rayner, Todd and Sabin. They undertook this work from a hide which has since disappeared from the Naze. Little Terns, which nest at Hamford Water, are amongst the rarest sea birds in the country. They only nest on sand dunes and beaches and are therefore at acute risk from both disturbance from people and high tides. Weston was also extremely active in his opposition to 'development' (a term I personally dislike due to its suggestions of definitive progress) in the wider area, particularly in applications regarding coastal development. In one such objection he wrote:

I should like to suggest that we consider what our actions will do to the world we leave behind us. The pace of modern development means that the complex web of interrelated natural species does not have time to adjust to changes in the environment. Can we with a clear conscience exploit and destroy habitats for short term profit and pleasure, and leave a drastically impoverished environment to future generations.[200]

Environmental affairs are seen politically as a 'niche' issue but the environment has no boundaries politically and encroaching development impacts in so many ways, in terms of landscape and habitat for non-human animals and plants but also critically in terms of social welfare, mental health and community value. In nature conservation terms it is vital that the Essex coast retains its stretches of undeveloped coastline. It is one of the few counties with large areas of saltings and estuaries. The saltmarsh in Hamford Water is built up slowly over long periods of time and the roots of the vegetation help to stabilise the mud, providing a vital ecosystem service. Hamford Water is home to sea purslane, cord-grass, sea aster, sea pinks and sea lavender, among other plants, and a wide variety of invertebrates, many of which burrow into the mud around the saltings. These include polychaete worms, bivalve molluscs and small crustaceans. Salt marsh is a precarious habitat and is critically vulnerable to sea level rise. Hamford Water has international importance as a wetland habitat for numerous nesting and migratory birds, many of which breed on the salt marshes and dunes. Redshank, Ringed Plover, Oystercatcher, Little Tern, Common Tern, Shelduck, Lapwing and Black-headed Gull all breed at Hamford Water. All of this goes on adjacent to an area with an increasingly urban identity and the busiest container port in the country. Today, wildfowlers are some of the most dedicated and interested conservationists. However, this has not always been the case, most notably exemplified by the thirty foot steam launch *Tyche*, used by Colonel Davis of New House Farm in the 1890s. The *Tyche* had two swivelling fowling guns mounted forward, one of them a two inch breech loader and the other a one and a quarter inch muzzle loader.[201]

Hamford Water is also known as the 'Backwaters' or the 'Walton Backwaters' and has inspired a number of authors, perhaps most famously Arthur Ransome who wrote *Secret Water*, the eighth book of his *Swallows and Amazons* series of books for children, published in November 1939. The Backwaters were also immortalised in the 1971 BBC adaptation of *The Snow Goose*, based on the 1940 short novel of the same name written by the American author Paul Gallico. The film was directed by Patrick Garland, starred Richard Harris and Jennie Agutter and was shot mainly in Kirby-le-Soken and Landermere. It tells the story of Philip Rhayader, an artist and conservationist and a local girl called Fritha who spark an unlikely friendship as a result of a shared experience with a goose, wounded by a gunshot. The simplicity and fraternity of this relationship is set against the horrors of the Second World War.

In addition to the creation of the Essex Wildlife Trust reserve, 1970 was a key year in the Naze's history because of the opening of the water treatment works at the northern edge of the Naze, built on what used to be 'New Zealand' field. The treatment facility was much needed at the time due to the rapidly growing local population and facilities that were hopelessly out of date. The scheme took decades to come in to operation, having been first proposed in 1935. The facility contained high rate bacteria beds and employed a short outfall in to the sea, avoiding pollution of the beach. The works was designed to serve a population of 18,000 across Walton, Frinton, Kirby-le-Soken, Kirby Cross and Great Holland, with an allowance for future extension to serve double this figure. Construction began in May 1966 and it took nearly 8 years to complete over a number of stages, eventually costing £1,350,000. In 1995 Barbara Oxborrow recounted her memories of the scheme:

> It was a great day on 14th May 1970 when the large and complex scheme, funded by a comparatively small authority, was officially opened. The cavalcade, led by the Chairman's old Bentley, followed by the vice-Chairman of Essex County Council's new Bentley, followed by the cars of the councillors, officers of the council, contractors and invited guests, set off from the council offices in Old Road, Frinton and proceeded, by way of the pumping station on the Bathhouse Meadow in Walton to the treatment plant at the Naze. After the obligatory 'few words', the Chairman was present with, and drank, a glass of sparkling water from the outfall, after which all repaired to the Grand Hotel in Frinton for lunch, at which many more words were spoken.[202]

Once a sparsely populated backwater, the area surrounding Walton and Hamford Water has become increasingly urban both in landscape and character. With the prospect of yet more development in the future it is vital that planners remember to account for the value of leaving places such as the Naze 'wild', not just for wildlife but for local people. History that dictates future action is not often good history. People in the past should not be judged for their actions using present perspectives or ideas. The present should similarly not be judged on past principles or observations. Nonetheless if the pace of change is so great to affect an area's wellbeing then history should be used to reflect upon where an area is heading, in terms of its identity. We face a tipping point in deciding whether to grow the built area further or to hold the line. 'Penurbia', a term used more in the United States than in Britain, signifies a landscape or community whereby the character of a place is situated in the liminal space between urban, rural and suburban. This is where southern and eastern parts of Hamford Water are increasingly placed. A future for the Naze needs to set against a discussion regarding how people wish the future of the area as a whole to look and feel. On the Naze a number of land-uses reflect the area on a microcosmic level. To the south and east lies both residential and industrial built development, to the north and west there is the Walton Hall farm estate consisting of arable fields, grazing marshes and beaches and to the east and north east there is the public open space and nature reserve, containing a number of different habitats and places for recreation. Environmental stewardship has become increasingly important at Walton Hall farm in an attempt to safeguard the environmental value of the space. However, with sea level rise and erosion to the east and north east the farm is increasingly 'threatened' by nature. It will have to move with the times and adapt if it is to continue as a viable working farm, an identity that it has clung on to in the face of change, for over a thousand years.

CHAPTER 5

Walton Hall House and Curia

My own childhood memories of Walton Hall house are of it in a redundant, crumbling state, a house in serious need of care and attention but one that was a perfect host for the imagination to run wild. For most of the twentieth century the hall house stood as a derelict structure, waiting for a new lease of life having lost its role as the principal farmhouse for the Walton Hall estate before the Second World War. As an eight year old looking up at the building in the 1990s it seemed more like a fort than a house, the square tower block standing steadfast, determined not to crumble completely away and yet it seemed to cry out for someone to reach out and help it. Sometimes the wind, famously blustery at the Naze, would whistle through the timbers and windows in such a way that it sounded like someone was calling out for help. It was a ruin, and not a very romantic one at that, and one was able to see through all the floors from bottom to roof. This all changed at the beginning of last decade when it was sold to a developer who restored the building and for the first time in nearly a century it became a liveable home and entered a new chapter in its long history, albeit divorced from the farm it served for centuries.

As the traditional manor house of the Walton Hall estate (although in reality a simple farmhouse or steward's dwelling), Walton Hall has undergone several redevelopments, reuses and replacements over time. There is documentation to show that there has been a manor house in Walton since 1243, although it is assumed that the curia and the house existed from before this time.[203] The present Hall is not the 'original' house but has been on its site for several centuries at least. The aforementioned 1243 reference is to a lease document of Canon Richard de Staunford which mentions a 'bona aula' or literally a 'fine Hall' but one that had apparently been in existence for a significant time beforehand, suggested by the alternative description from 1243 of the existence of 'an *old hall*'.[204]

John Button has undertaken substantial research on the medieval demesne buildings and I acknowledge the extent of his work below. He implies that at the point in which he is assessing, in 1297, the Hall had a purely functional purpose.[205] The main house was used to entertain dozens of people, whether for church business, ceremonial occasions, feasts or other social events. From de Staunford's lease document it is shown that there was a large open 'barn-like' space at the centre of the thirteenth century complex, used to accommodate many people. Attached to this building was a 'dispensa', which was an office for the steward, a cellar, a bedchamber and a room called a 'solar' which let light in to the building. Additionally there was a large chamber to the west of the Hall and a small chamber to the north, both connected to the central room with covered passages. There was a large kitchen in which there was a bakery, a brewery and a wide oven, capable of catering for the many guests the lord of the manor entertained whilst in residence.[206]

By 1251, it is noted in the lease of John de Cheam, that a malting house with a thatched roof (*torallium coopertam*) had been built, complementing the existing brewery that had existed for possibly a hundred years beforehand.[207] Ale was a staple drink in the medieval age for rich and poor alike, possibly due to the difficulty in finding sufficiently clean drinking water but also because intoxication is a timeless form of entertainment and release! Barley was one of the principal crops of the demesne and so the brewery and maltings were in operation throughout the year. However, certainly the drinking of ale was often a safer exploit for the villeins than for their masters, as they made use of earthenware and wooden cups, whereas the elite used utensils made of poisonous lead.

In the Cheam lease entry of 1251 it is written that a 'capella', meaning chapel, had been built inside the curia, with five glass windows, a sign of significant wealth and prosperity.[208] The lord of the manor was often a senior member of the Cathedral and therefore the existence of a magnificent chapel was thought suitable. In a lay manor house a chapel this splendid would have been rare. Richard Rufus, who held the lease in 1150, was an Archdeacon of Essex and Richard de Staunton was a canon of St Paul's. The chapel was for the private use of the lord but worship was also available for the common people, not within the curia chapel, but in the parish church. When the lord was away, the parish vicars, often villeins themselves, continued to lead worship and the leadership of the estate was taken in charge by the estate steward.[209]

Within the curia, which was surrounded by a wooden palisade, were a variety of ancillary farm buildings. These included, in 1243, two dairies, a small stable, a large granary, known as 'the Great Barn of Walton', a small granary and two hen houses. There was also a larger stable outside of the palisade wall and another seven barns scattered around the demesne, some possibly on the marshes for sheep.[210] In 1251 it is documented that there was also a dovecot, a new stable and a small barn for 'karros et carrettos' (wagons and carts').[211] The tenants of Kirby and Horlock, who did not have the use of any farm buildings at the Kirby curia, were expected to maintain

101

the woodwork of the barns at the Naze as well as the surrounding palisade wall, as was necessary, in return for making use of the barns themselves.[212] It is not known where the Kirby tenants stored their produce, although the amount of sheaves described as being stored in the 'Great Barn' when Richard Ruffus took over (c.1150) would suggest that the barn was used for storage of additional produce outside of the Walton demesne.

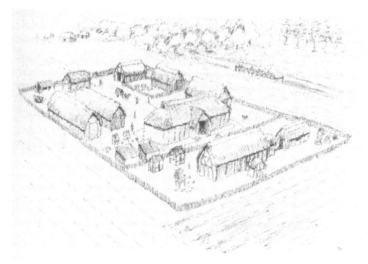

Above: Sketch of a possible viewpoint of Walton Hall curia circa 1297. Courtesy of John Button.

The functionality of the curia was of primary importance in how it was designed (refer back to chapter three to see the positioning of the curia in relation to the demesne). The site of the Walton curia was moved inland as the sea encroached upon it. By 1304 it is clear that the curia was under threat, as the great stable, which was placed just outside of it, was being used as a temporary cow house because the latter, as well as several other buildings placed elsewhere on the demesne, had been swept away by the rapidly advancing sea.[213] In 1335 a visitation of the Chapter of St Paul's followed with certain observations suggesting a change in circumstances for the manor:

> At Walton the chapel, the barn close to the chapel, the solar close to the inner door by the same adjoining barn, the old kitchen and the covered passage between the said kitchen and the hall, and one barn which a certain baliff, Elkin by name, built are in a state of total disuse and would be expensive to maintain.[214]

By the mid fifteenth century the medieval curia had been abandoned to the sea, with the old Hall converted into another temporary barn. The new Hall and ancillary buildings had been built by 1458, noted in the lease of Canon William Radcliff, although again it is not known whether these were situated at the current position of Walton Hall farm.[215]

In 1544, St Paul's disposed of the estate as part of the great transfer of the Sokens with Crown estates further south and west in Essex, Hertfordshire and Middlesex. During the lordships of the Darcys and Rochfords the house was occupied by an estate steward, appointed to manage the affairs of the Naze and Walton on behalf of the lord of the manor. John Norden records in 1594 the existence of a great brick tower attached to the house as defence against 'sea robbers' and describes the house as being 'of no account' otherwise.[216] He wrote of 'Walton Hall, a house of noe accompt (sic), a ferme; but to be remembered, in regarde of a stronge towre of brick for defence against sea robbers.'[217] The reference to 'sea robbers' could be to Barbary pirates who originated from the North African coast and famously stole both possessions and people to make slaves from coastal communities in north-western Europe. By the reign of James I they were becoming a serious menace along the Iberian, Biscayan and British coasts. It could also refer to smugglers attempting to avoid customs by landing in the Backwaters. It can be assumed that the block tower part of the current house is what Norden was referring to. The door was placed on the first floor meaning that entry to the building could only take place by erecting a ladder from the outside. In a mid-nineteenth century advertising booklet for enticing wealthy London folk to Walton, it is written that'...a cheerful, garrulous old lady told us that in the last century the Hall was used as a prison, principally for smugglers.'[218] However, this is as close to evidence that we can achieve. The distant memories of a ninety year old woman, gathered through hearsay and added to an advertising booklet, of which it would be in the interest of the editors to dramatise any story to entice visitors to the town, are sadly not the best grounds for evidence for the building being used for such a purpose.

It is disagreed exactly when the 'block tower' part of the current house was restored as a functional part of the Hall house although it would make sense of Norden's account if the current 'block' structure had been in place for many centuries with a separate Hall dwelling built nearby and later the two buildings were joined. Captain Greenville Collins's navigational chart of 1686 instructed sailors to steer by the Naze, 'which may be known by the tree and the house that stand on it'.[219] 1686 was before the building of the Naze Tower and therefore the house would have been the most prominent building, albeit in addition to a small wooden structure, a predecessor of the Naze Tower. In Collins' account the house was reported as being a forty foot tower, which would make sense of Norden's map. John Button believes this was demolished and rebuilt in the eighteenth century.[220] If this was the case then it is likely that the Hall has been in its present position since the early seventeenth century at the latest. If it can also be linked to the 'new hall' of the William Radcliff lease of 1458, then the Hall has been in its position since the fifteenth century, perhaps based upon an even earlier structure.[221] It seems that the position of the curia remained constant from 1150 to 1468 and that Radcliff's mention of the 'new hall' had been built near to the curia complex.[222]

Before the building was restored in 2002-2005 a short survey was made and it was discovered that the bricks varied in age. There were courses of what appeared to be 'shallow irregularly shaped Tudor bricks' laid over 'standard sized eighteenth century bricks'.[223] The structure of the cellars was also noted to date from around the sixteenth century, the fireplace especially being recognisably Tudor in style.[224] We could assume therefore that the present house is, at the least, based upon the foundations of an earlier structure. This early structure could date as far back as the fifteenth century, referencing the Radcliff lease of 1458, which denotes that a 'new hall had recently been built on the Naze' and that it was 'close to much older farm buildings, one of which was a granary apparently converted from the fabric of an older hall'.[225] This suggests the current site of the barns is many centuries in age.

Walton Hall has attracted many story tellers, the most famous of whom is probably Ursula Bloom who wrote 'The House that Died Alone', based directly on Walton Hall and published in 1964. Bloom was a local lady who profoundly disagreed with Godfrey Evans' development proposals for the Naze and knew Walton Hall well. The novel, which is set across three different time periods, is partly a commentary on the development threats the Naze was facing and partly a response to the conversations she had with those who knew the house well and lived in it before it became a ruin, which, certainly by the 1960s it was well on its way to becoming. Although Bloom makes it clear that the story is entirely fictional in a note beforehand, she presents the house exactly as many people saw it, as well as commenting on the demise of Delstone

(Walton) which many people felt was happening. She also wrote of the self-important self-appointed status of Pratwell (based on perceptions of Frinton). Bloom was a talented writer and captured the essence of the Hall so well that for many, the ghost story she presents in *The House that Died Alone* has become almost fact, even if most have never read her novel. Her opening words ring true with the experiences of many (myself included) who grew up knowing Walton Hall as a ruin:

> The house stood dying...its windows had gone, the wormy frames stolen for firewood by marauders, and there remained only the jagged stone outline of where once those windows had been. The stairs had collapsed, sinking down and spitting dust, splintered balustrade, and cobweb all in one. The lead had been stolen from the flat roof, which was now mercilessly bared to the stars, the hurricanes and the rain. But the defiant house stood there, determined to die as it had lived, proudly, even if to the sound of a woman weeping.[226]

The sale particulars of 1801 presented 'the tower' as a completely separate residence, a 'shooting seat intirely (sic) independant (sic) of the Homestall'.[227] 'The Tower' is said to be 'near to the Farm House, but intirely (sic) independant (sic) of it...completely fitted up and augmented by the late Proprietor, at a very considerable Expence, not less than £1500 and made a compleat (sic) shooting seat, with Kitchen, Cellars and Servants Rooms, fit for a small genteel family, both for a Summer Residence, and a sporting seat in the Winter, and is compleatly (sic) furnished.' Although the reference to 'tower' could be to 'the Naze Tower', it is very unlikely. The specific mention of 'cellars' involved directly in this account, suggests it is the block tower.[228] From this we learn that by 1801 the 'tower' was not part of a single Walton Hall farmhouse but a separate residence entirely. In the 1844 survey of Walton Hall, the surveyor, G.R. Jay of Great Bromley, illustrates a building similar to the current structure both in appearance and in architectural value suggesting that between the 1801 sale following Rigby's death and 1844 the house was transformed.[229] There also appears to be a greater number of different barns in the farmyard in 1844.

Above: Walton Hall. Photograph courtesy of Putmans Photographers - The Walton Archive – www.putmans.co.uk

The 'tower' itself is a square, rendered, brick building set over three floors. Whatever the origins of the building, by 1827 the Hicks family were using it as a 'shooting seat' as it had been used in the late eighteenth century.[230] A 'shooting seat' is referred to in the 1820 valuations of the estate – 'the Tower, now a shooting seat to the value of about £500' as well as in the 1801 sale particulars.[231] Unfortunately the 1820 valuation does not suggest whether the tower and the farmhouse had been joined by that time. However, the 'tower' was a domestic part of the main house by the mid nineteenth century at the latest, accommodating the formal rooms used to entertain guests and servants quarters above. It was rebuilt when the house was restored in 2002-2005, working to plans from nineteenth century evidence of the building.

Certainly, the barn complex that is currently situated at Walton Hall farm, opposite the Hall house, has been in place since John Hicks arrived at Walton Hall as lessee. The main barn and the adjacent smaller barns were built by Barnard or Hicks circa 1820, although some of the existing structures predate both Hicks and Barnard and were in place before the sale to Rigby in 1775. These include the small granary to the west of the barn complex, the timbers of which were reused from older barns, possibly dating to the time of the medieval curia. It is difficult to precisely date the structures in place at the barn complex due to the apparent lack of clarity of contemporary maps and lack of documentation of the seventeenth and eighteenth century farm buildings. Andrews and Drury undertook the first large scale map of Essex in 1776, to little over one inch to the mile, and although buildings are shown where the farm is now situated, it is not clear what these buildings are. Chapman and André's map of 1777, drawn to two inches to the mile, is similarly vague.

The barns were in agricultural use for the entirety of the nineteenth century and the majority of the twentieth. The main barn was used first as a threshing barn and then as a milking shed up until the late 1960s when Eagle's herds were moved and combined at Brick Barn Farm, Kirby-le-Soken. The cart lodges were then used for young bovine stock until operations on the farm were contracted and the business became an arable enterprise in the late 1990s, reflecting the shift across the area from dairy to arable. The barns are listed grade two. However, over the years, most have fallen into disrepair, as the house did before them, and the eastern courtyard of buildings burnt to the ground following a fire in August 2013, destroying with it a significant private collection of munitions from across Russia and Europe [232]

Above: Walton Hall in the late 1880s. Photograph courtesy of Putmans Photographers - The Walton Archive – www.putmans.co.uk

The photograph shown above, taken at the turn of the twentieth century, illustrates how the Hall was very much a working farmhouse as well as a home for the farmers that worked the land. The structure at the very back of the Hall house was used as the farm office. J.W. Eagle lived in the Hall and worked the land during the ownership of Sir John Johnson in the later part of the 19th Century. His son, another J.W. Eagle, was able to purchase the house along with the entire Walton estate in 1920. His son in turn, my grandfather, yet another John Woodruffe Eagle, wrote about his memories of life at the Hall as a child:

> My earliest childhood memory is of a large garden and a great grey house overlooking the Essex marshes and the sea. The rambling garden, the boundary of which contained my familiar world, seemed itself a vast country filled with hidden paths and the most wonderful secret places. It was a land of enchantment for a rather solitary small boy, who had many hours in which to find his own amusement.
>
> The house was very different. It was also vast and rambling but, except for my own domain of day and night nurseries at the very top, the rest filled me with dread. It was a house full of dark stairs and passages, places through which I would run as if the 'Erl King' was behind me and huge cupboards which yawned like the abyss. However, although the house was gloomy, my days were happy enough. I was used to amusing myself and I had all the toys that a small boy could desire. I spent long days in a private world of my own devising, where my favourite toys were people of the greatest importance, and I ruled supreme. No state endured more cruel wars nor won more shining victories than mine on the nursery floor.[233]

My grandfather's memories seem to emit a mixed response of both respect and uneasiness. In fact, when I have spoken to anyone who knew it as a place to live or visit, it is this mixed response that I have received. A.H.B. Franklin, the goddaughter of Dorothy Eagle (my grandfather's mother), wrote in 2005 of her memories of visiting Walton Hall. I have quoted her at length as I feel she gives a worthy insight to how the Hall actually 'felt' before the war:

On arrival at Walton-on-the-Naze we were met and driven through the village and up the long hill to the Naze in a yellow wheeled dog cart. If we arrived in daylight, we might catch sight of Grimes or Emma, the peacock and peahen named after the famous Colchester tramps. The house on approach was a rather gaunt, grey stucco square building with an elegant Georgian entrance and large downstairs windows reaching to the ground.

As far as I can recall, the drawing room was a typical example of my Godmother's artistic sense; the colours, perhaps largely lilac and green; there were alcoves in the walls for ornamental porcelain but above all, the smell – the perpetual smell of Turkish cigarettes. The dining room was masculine and practical, with a huge sideboard holding numerous silver trophies awarded for the Eagle's prize winning flocks of sheep.

...Further rooms downstairs are beyond recall, apart from dark passages especially to the kitchen. The oldest part of the house, it was a hive of store rooms, larders, boot cupboards, racks and shelves. There was a good main stair case. I cannot remember if it was circular or dog legged. I am certain the landing was the former and the doors leading off it were curved.

...A long dark corridor running off the first floor landing was impeded in one place by a large beam running across it. It was a hazard over which I tripped and fell on several occasions. However, beyond this obstacle was "the Den", my Godmother's chintzy and comfortable private sitting room. Strictly by invitation, on entering I was always offered a dish of pastel coloured, sugared almonds.

I had a familiar pretty bedroom, overlooking the garden; unblemished by hints of haunting...As far as I am aware, it [Walton Hall] has always had a reputation for a mysterious "something". One legend records a Lord Devereux shooting himself in an upstairs room after losing a fortune at the card table; another that local prisoners were locked in the extensive cellars until brought to trial, creating an almost tangible atmosphere of human misery.

...The only thing that marred my visits was bath time every evening. Not an aversion to soap and hot water but the long journey up a dark back staircase to a gloomy, pitch pine panelled room with an enormous bath.[234]

Her detail is glorious and effectively illustrates the feeling of mystery about the house, emanating Ursula Bloom's fictitious account. However, the section apropos the haunting can probably be put down to after dinner story telling. The Eagle link with Devereux farm and the name of the mysterious committer of suicide surely cannot be coincidence. Nevertheless, there was evidently a gross feeling of uneasiness for several associated with the house, something that Bloom was keen to capitalise on in *The House that Died Alone*.

Eagle senior continued to live at the Hall with Dorothy his wife and Frank ('Bunk') his brother until 1928. The reason for their moving is supposedly down to Dorothy, although the economic depression of the 1930s had an equal part to play. On the one hand, it was not financially viable for the family to remain living in such a building. However, just as important was Dorothy's disquiet about the place. She was very conscious that 'something' was 'amiss' at the Hall and persuaded Woodruffe to leave. They moved instead to Walton Ashes, a farmhouse on the Kirby/Walton border.

The Hall was let to a Colonel named Felton, a celebrated florist in Knightsbridge and his wife. He was, according to A.H.B Franklin, a 'keen researcher in to the paranormal' and so relished the opportunity to live in the house.[235] His wife again had troubles with the place and after not very much time they left. The house and estate, along with the rest of the Naze, including the golf club, were taken over by the army as a training camp during World War Two. The house fell into complete disrepair following the exit of the armed forces. JW Eagle considered returning to the Hall following the war but the house was no longer in a habitable state and was by no means a safe structure in which to reside. It seems that Wooodruffe was particularly aggrieved however that, upon his return, he found that the entire wine cellar had been emptied. The family took the decision to abandon the place, indefinitely remaining at Walton Ashes. For the Eagle family feelings towards Walton Hall had never been one of contentment, to say the least, although it seems sad that the house was left to 'die alone', to use Ursula Bloom's phrase.

A.H.B. Franklin revisited the house one evening when staying with her mother in Frinton in March 1946, writing later of the devastation that the army had caused during the war:

> I was horrified to see what had happened to the house and grounds. A drawing room window was open and I decided to go inside, taking my dogs with me. But nothing on earth would induce them to cross the sill. So, I tied them up and went inside, although it was getting towards dusk. The damage, decay and endless graffiti brought me close to tears.[236]

After many years of dereliction the Hall house was sold by the Eagle family in 2002 and it underwent full restoration which was completed in 2005.

Walton Hall house and the farm buildings are not the only buildings at Walton Hall farm. On 27th March 1903 Sir John Johnson sold 1 acre of Groveshot field to the Admiralty so that they could construct a station to act as a base for the coastguards. 'The Coastguard Cottages', or 'Creek Cottages' as they are called today, were built so that the coastguards could easily access the station, which was situated at Stone Point at the beginning of the 20th Century. They were designed for four families and a single man who would keep watch in the lookout on the northern edge of the terrace block. The cottages were in use by the coastguards from circa 1903-1906 when the station was moved to its current location closer to the town. The Admiralty sold the cottages, along with the acre of Groveshot purchased from Sir John in 1903, back to the Hall estate in October 1923 and they were renamed 'Creek Cottages'.[237] They were used as farm cottages for much of the twentieth century, and provided accommodation for the men who worked on the estate as well as for their families. Mike Pratt was the son of the gamekeeper, Bill Pratt, and grew up at Creek Cottages. His detailed account of life on the Walton Hall estate is set out in the appendix (oral history interview with Mike Pratt). The other set of cottages at Walton Hall are 'Walton Hall Cottages', the distinctive yellow buildings that can be seen clearly from as far as Beaumont, set against the white of Walton Hall and the red roofs of the barns. The cottages date from circa 1910 and were originally used as accommodation for labourers.

When sources are rare it is stories that tend to prevail. This is certainly the case with Walton Hall which has (although perhaps less so now it is no longer in a ruinous state) inspired many stories. Once standing on the peninsula with few neighbours other than the Naze Tower and the few farm cottages dotted around the Naze, development has now encroached upon it and, more broadly, the Naze part of Walton is now as urban as the town. Norden wrote of it being 'of no account' but certainly the mysteries that surround it mean that it is of anything but. That bricks and mortar can have an identity is for some absurd but certainly the connotations Walton Hall has attracted, both in terms of mysterious fiction and perilous fact, have given the house a defined identity that relies just as much on the unknown as the known. I set out in this research process to look behind the stories and discover some facts behind the architectural and cultural history of the building. However, I have discovered that it is not possible with the little evidence available and that in reality it is the mystery and intrigue surrounding it that make Walton Hall such a fascinating building and the stories themselves are an intrinsic part of the building's history.

CHAPTER 6

The Naze Tower

The Tower on the Naze, standing at eighty-six feet high, is a spectacular feature and vivid landmark on what is otherwise a mostly flat landscape with the 'big skies' for which Essex is famous. The Naze Tower can be seen both from the sea and from far inland, and it is an important piece of the Naze historical and cultural jigsaw, itself a key part of the Naze's maritime and cultural history. It was granted Grade II* listed building status by English Heritage in 1984 due to its special architectural and historic interest. As marked on the building, the maritime corporation of Trinity House commissioned the Tower in 1720. It was designed by William Ogbourne, built under the supervision of Daniel Fisher, completed in October 1721 and cost £784.[238] The Tower can be seen for many miles out to sea and prior to modern navigational technology served a vital purpose in guiding vessels through Goldmer Gat, north of Gunfleet Sands and immediately opposite the Naze.[239] The Tower stands near the old 'Nase' point, at the line of the nose of the promontory, the vast majority of which has eroded since the early eighteenth century.

The building was not the first construction on the Naze to mark the peninsula. The problem for the historian comes when deciphering from documents the usage of the word 'tower' for it appears on maps many years before 1720. In some cases, "Walton Tower" is clearly marked over a hundred years beforehand. Early maps of Essex are extremely vague in their constitution and are not usually drawn to scale. However, marked features can provide useful information to put the changing nature of landscapes and places in to context. In his map of Essex in 1576, Saxton specifically shows and names "Walton Tower" indicating a structure to mark the promontory. Maps from Norden (1594), Speed (1610), Jansson (1636), Blaeu (1645), Jansson (1646) and Speed (1676) all name and mark it. Ogilby and Morgan, in their map of 1678, go further and give more detail, showing a symbol designating a 'beacon' where the current Tower is situated. From 1726 onward the Tower is marked as both a building and a beacon, indicating the newly built brick structure. Warburton (1726), Bowen (1749) and Gibson and Bowen (1762) all mark it as "A Sea Mark" with a lighthouse symbol. G.R. Jay, the surveyor of Great Bromley who drew the estate plan of 1844, also marks the Tower as being "A Sea Mark".[240] The Essex historian, the Reverend Philip Morant, who wrote *The History and Antiquities of the County of Essex*, published in 1768, mentioned 'the building of a new eighty foot brick tower by Trinity House' at the Naze.[241] The word 'new' is used loosely in this context as the building had been a prominent part of the skyline for over forty years previously.

Despite the inscription and evidence pointing to the construction of the 'Naze Tower' in 1720, it is clear that 'Walton Tower' was a significant feature on the north Essex coastline perhaps more than two hundred years previous to the building of the brick structure. It is probable that the 'Tower' mentioned prior to 1720 was a simple wooden beacon tower of about thirty to forty feet high, used to mark the promontory and send warnings up and down the coastline. This is supported by evidence from the 1678 map from Ogilby and Morgan which shows that near identical 'beacons' were situated elsewhere, such as at Harwich, St Osyth and Foulness. It is likely that it was felt by Trinity House in the early eighteenth century that the existing system was not fit for their purposes, hence the building of the current structure. Harwich, home to a naval base since the 1660s, had a busy harbour and according to Daniel Defoe who visited in 1722, just a year after building works at the Naze Tower were completed, Harwich was 'able to receive the biggest ships of war and the greatest number that ever the world saw together'.[242] Investment in the surrounding area of Harwich was therefore required to ensure safe passage for vessels moving in and out of the harbour. The earlier 'tower' structure stood at the Naze for a hundred years after it became redundant, finally being taken down in 1824.[243]

Shortly before visiting Harwich Defoe travelled to Walton, where he made reference to the newly constructed tower and the erosion experienced at the Naze. He wrote that:

> on the north-foreland (of Walton) is a new mark, erected by the Trinity-House men, at the publick Expence (sic), being a round Brick Tower, near 80 Feet high. The Sea gains so much upon the land here, by the continual Winds at south-west, that in the memory of some of the Inhabitants, above 30 Acres of land have been lost in one Place.[244]

At this point the Naze Tower stood relatively far in land, in contrast to its current position just fifty metres from the edge of the cliff. When the Naze Tower was built, Trinity House held sole responsibility for the safety of all marine vessels and maritime activity surrounding the British coastline. The body was established in 1514, upon the grant of a Royal Charter by Henry VIII to the 'Guild of the Holy Trinity' although its origins date back far previous to this date. In 1566, Trinity House was given powers under the Seamarks Act to build 'So many beacons, marks and signs for the sea, whereby the dangers may be avoided and escaped and ships the better come in to their ports without peril.'[245] As funds grew, under the bolster of profit from the provision of ballastage from dredged material, increasing numbers of 'sea marks' were built, although none as prominent or architecturally complex as the Naze Tower. The growth of the port of Harwich meant it was critical to mark the promontory of the Naze to avoid ships grounding ashore, whilst navigating through the sand banks. However, the Tower served no other purpose for Trinity House other than to act as a marker. The internal space of the structure was therefore largely irrelevant to the organisation and so the lord of the manor of the Walton Hall estate generally acted as 'custodian' of the building.

Above: A 1742 etching of Walton and the Naze Tower standing independently on the promontory. Courtesy of Putmans Photographers - The Walton Archive – www.putmans.co.uk

During Richard Rigby's ownership of the estate in the late eighteenth century, permission was gained from Trinity House to hold exclusive 'tea parties' in the Tower for the political and social elite of the day. Rigby was a friend of John Montagu, the Fourth Earl of Sandwich, a fellow Bedfordite and the man who is first said to have put a piece of beef between two slices of bread and 'invented' the sandwich, who, at the time, was having an affair with the singer Martha Ray.[246] It is thought that Sandwich attended the parties with Miss Ray. Miss Ray had lived with the Earl as his mistress since the age of seventeen. Sandwich was married but separated from his wife who lived in Windsor and suffered from 'mental illness'. It is thought that the parties at the Tower were raucous affairs, often lasting long in to the night with much more than tea being consumed.

In 1775 Martha Ray met James Hackman, then an army lieutenant, at Lord Sandwich's house at Hinchingbrooke. Their friendship developed and Hackman became besotted with Ray although it is disputed whether they were lovers at any point. Ultimately however nothing came of it but Hackman remained highly jealous of any 'rival' suitor. On 1st March 1779 he was instituted as Rector of the parish of Wiveton in Norfolk. A few weeks afterwards on 7th April 1779, Martha Ray went with her friend Caterina Galli to see Isaac Bickerstaffe's comic opera 'Love in a Village' at Covent Garden Theatre.[247] Hackman had followed her there and suspected that she had a new lover. After the performance, Miss Ray stepped out of the theatre in to the road. The clergyman stepped towards her, shot her directly in the temple and she died on the spot. Hackman was tried for murder and hanged at Tyburn on 19th April. The parties at the Tower stopped soon afterwards. When Rigby died in substantial debt and the estate passed into the hands of a whole series of local farmers and officials, the Tower became neglected.

Although not designated a Martello tower, and should certainly not be called as such, the Naze Tower was used as a signal and watch post during the Napoleonic wars at the turn of the eighteenth century.[248] Interestingly, during the Second World War, the Air Ministry mistakenly called the Tower a Martello Tower despite the inscription on the front of the building: 'TRINITY HOUSE 1720', indicating a construction well before even the birth of Napoleon Bonaparte. In Edwardian England, the Royal Navy used signalling flags and a semaphore post at the Tower to practice manoeuvres and communicate with ships and other signal posts along the coast.

The Tower was sometimes used as a Victorian tourist attraction, for those who were sufficiently interested. In his 'descriptive account of Walton-on-the-Naze', published in 1860, Thomas Wilmshurst notes that:

> From the summit of this tower (which is accessible to the visitor, through the courtesy of J.W.Hicks Esq. of Walton Hall, who is the custodian of the building) an extensive prospect is obtained of the German Ocean, Harwich, Landguard Fort, and the Suffolk coast; and on a clear day the distance to which the eye can reach is most interesting, the Kentish coast being certainly discernible: and the view inland completing a panoramic picture, which well repays the labour of the ascent.[249]

The First World War brought a return to its role as a lookout post, with a beacon added and sentry boxes put in place on the ground beneath. In the 1930s it was used by the Naze Golf Club as a 'ground shelter', following an agreement between the Club and Trinity House.[250] During the Second World War, Royal Air Force operators were stationed at the Tower and a Chain Home Low Radar Dish was installed on the roof in 1942 to detect enemy aircraft and ships. A fascinating account of the Tower's history during this period has been written by Reg O'Neil MBE and explains from primary experiences the realities of war on the Essex coastline and the role that the RAF played in the Tower's diverse history.[20/251] The other publication to discuss the period is Geoff Rayner's *Seaside Front Line*, which, in diary form sets out the wartime

---

[20] I highly recommend this book if you wish to explore this period in detail.

activities on this stretch of the coast. It appears that the Tower came very close to destruction during the Second World War as, in a series of recommendation on local defence in August 1940, the Commanding Officer of the local infantry battalion suggested that the Naze Tower should be demolished as it formed a good landmark and thus could assist enemy operations. This followed about fifteen high-explosive bombs and two incendiary bombs falling on the Naze golf course on Monday 19th August 1940, close to where searchlights were placed.[252] Bombs continued to be dropped on the Naze throughout 1940. On 31st August of that year a British aircraft was shot down, forcing the pilot, twenty four year old Pilot Officer Gerard Maffett, to bail out at about four hundred feet, far too low for his parachute to take effect and sure enough his body was found on the sand by the Tamarisk seawall, just yards away from the wreckage of his Hurricane.[253] The remains of this Hurricane were recovered in 1973 and are now presented at the Battle of Britain Hall of the RAF Museum at Hendon. The other two crashes to occur at the Naze included a Hurricane MkIIC, Z3064, from No.3 Squadron, at Stone Point on 29th April 1941, killing twenty year old pilot Sergeant Richard Brewin and the incident in March 1945 when seven Canadians and one Briton, died when Halifax MkVII RG475 crashed on Walton Hall marshes. Flight Lieutenant John G Clothier was buried at Pentwyn, Wales with the Canadians – Squadron Leader Edwin Hayes, F/O Colin Hay, P/O Joseph Ringrose, F/Lt Marius Nielson, F/Sgt Gilbert Orser and Sergeant Douglas Cooke are together at the Commonwealth War Graves Cemetery in Brockwood, Surrey.[254]

In the late 1950s, as the Cold War set in, the United States Air Force took the Tower over in their efforts to intercept Soviet information. Detachment 4 of 603rd Communications Squadron, whose headquarters were at Weathersfield, Essex, was stationed at the Tower from March 1957 until June 1958 and unfortunately the building was not treated kindly. Afterwards, aerials were used by the Harwich Haven Port authority, the police and the coastguards but the Tower became derelict.[255] Trinity House sold the Tower in 1986. The asking price was £35,000. Roy Bradley, who lived in Clacton, was the first private owner of the building. Initially, he wanted to use it as a rehearsal space for his band, 'the Nitwits' but after finding it unsuitable planned to convert it in to a maritime museum for the town.[256] There was some local objection and surprise at the decision taken by Trinity House to sell the building. In a letter to the Frinton and Walton Gazette in 1986 J Clement Jones of South View Drive, Walton wrote:

> While not in any way questioning the good intentions of the purchaser of the Naze Tower, I share the sense of dismay and disappointment of many at the manner in which Trinity House disposed of this very significant landmark. I feel that a national body of the tradition of public service and integrity of Trinity House should have handled the sale of the tower openly and announced to the public at large that it was for sale. The tower holds a very strong place in the affections of locals, visitors and yachtsmen alike. A Trinity House spokesman admitted in your paper last week that it is not their normal practice to sell off their premises that they no longer need in this fashion. He went further and said that they did not wish to be bothered by the very large number of enquiries they expected to receive. Surely this is condemnation enough out of their own mouths. If they anticipated considerable public interests, as indeed there would have been, should they not as an organisation of public repute have given the public, including organisations which might have been interested in raising the necessary

purchase money through public subscription, the opportunity of acquiring the Naze Tower.

Bradley's plans did not prevail however and the Tower remained derelict right up until the beginning of this century when it was renovated by the current owners, who purchased the building from Sven Bradley in 1996. The works required to the building were extensive. There had been severe water damage due to the trap door being kept open and the internal floors were dangerous. The upper two floors had to be replaced, new staircases added at the top (although the main Victorian staircase was in a sufficiently good state), the roof repaired and the castellation restored. It was eventually opened to the public in 2004 and is now a tea rooms and gallery and is popular with both locals and tourists.[257] Its establishment as a publicly orientated business has meant that, after so many years of closure to public eyes, a vastly important part of local and national heritage has been released, allowing people to enjoy this impressive structure whilst reflecting on its fascinating past.

The Tower is an emblem for the Naze and an attraction for Walton but it is also an important part of British maritime and commercial history. As a member of the Naze Heritage Project steering group the current owners of the Tower are working for a sustainable future for the landmark as part of the Naze. At the time of writing the Tower's owners are having it repointed and work is also being done to the roof to help secure its future. The building's 300th anniversary is now on the horizon and it stands, at the time of writing, just fifty metres from the edge of the cliff. Without the Crag Walk it is unlikely that the building would have lasted far beyond its 300th anniversary. However, gladly its future now looks secure.

CHAPTER 7

The Naze and the Sea

The cultural history of Walton has been shaped by the relationship between the land and the North Sea. Indeed, the whole of Essex's history is shaped by the sea. John Norden wrote in the sixteenth century that 'in the east Essex encountreth the mayne Ocean, an infalliable bounde'.[258] One of the most important things to realise when assessing coastal change is the enormous fluctuation that has occurred over time regarding sea levels. Throughout the earlier post-glacial period (the late Pleistocene/early Holocene), sea levels were much lower than today. It was during the Neolithic that the sea rose to roughly similar levels to the present day. The Bronze Age experienced a rise in sea levels that was again countered by a subsequent fall during the Iron Age. There was a rise during the Roman period but then a small fall in levels during the time of the Saxons. From circa AD 1250, the sea has been rising once again, accelerated recently by the melting of polar ice. There are several issues that must be covered when looking at the relationship between the Naze and the sea: firstly, erosion; secondly, offshore dredging, particularly with the close proximity of the Naze to the ports of Harwich and Felixstowe; thirdly, sea level rise and finally coastal defence. The rapidity of the coastal change that has taken place at the Naze is of particular importance in terms of the consequences it has for Walton.

Erosion at the Naze is noticeable in small time frames, due to the soft geological framework. Visitors who wait a few months or even weeks between their visits to the site will be met by a surprise that the coastline has visibly changed. This is mainly due to the 'slumping' of the cliff as several cubic metres will be undercut and then topple over the edge. Documentation of such land loss goes back a long time. The Reverend Philip Morant wrote how in the early part of the eighteenth century 'the raging sea keeps daily undermining and encroaching upon this parish, so that the Hall will soon be an island'.[259] Without sufficiently strong defences in front of and behind the Bathhouse Meadow in Walton town, this would again be the reality, albeit with much more built settlement on the Naze than in the time when Morant was writing. In 1798 the old parish church of Walton fell to the sea, kick-starting a new era of development in the town, based around the church of All Saints. The Naze used to project eastward and the 'nose' shape was far more prominent than it is today. Burnt House farmhouse, at the centre of the old farm that bore its name, fell in to the sea in December 1897.[260] High spring tides and winter storms have severely damaged or carried whole beach huts away in recent years, particularly near Green Lane at the Naze. These events act as a wake-up call, signalling the volatility of coastal settlement in the face of climate change, sea level rise and coastal flooding. The events of 1953 have now distanced themselves far into the history books for many people not to have it in the back of their minds. However, it is likely that we will experience a similar surge in the future. It is a question of when not if.

The East Anglian coastline has been eroding for the past eight hundred years.[261] The Naze is far more prone to the onslaughts of severe storms and surges than other areas of the coast, due to a peculiar combination of factors. Coastal erosion takes place whenever the sea is high enough to reach the base of the cliff, attacking it directly (corrasion or abrasion), causing pressure cracks in the cliff face (hydraulic action) and washing up debris that has already collapsed, which in turn collides

126

with the base of the cliff. The cliffs to the east of the peninsula have experienced large falls in the eastern cliffs, so much that the pill boxes that were constructed during the Second World War fell over the edge in the early 1960s and are now to be found forty or so metres from the cliff face, out on the foreshore. In 1941 these pillboxes were at least fifty feet behind the cliff edge, as shown in German reconnaissance photographs. Further, there is a problem regarding drainage and this also forms part of the reasoning for the rapidity of the erosion of the eastern cliff walls at the Naze. 'Slumping' is the result of rainwater percolating through the cliff and causing slippage once combined with the cliff being undermined by the energy of the sea.

A similar situation is found further up the East Anglian coast at the village of Happisburgh in Norfolk, which has been falling in to the sea for a number of years. Plans were drawn up in 2011 to 'resettle' Happisburgh although it differs to Walton in that houses are falling over the edge of the cliff today. The storm surges in the winter of 2013 resulted in several houses situated on the famous 'Beach Road' either falling over the edge or being demolished before suffering the same fate. For the Naze Tower, at least for the next century or so, the Crag Walk defences should protect the section of cliff that it lies directly above. The cliff will continue to retreat, before it reaches its natural angle of repose. However, Crag Walk may have to be lengthened in time to protect the remainder of the cliff face and to avoid outflanking, if this is the wish of future generations. Politically the situation would have been very different if Godfrey Evans had succeeded in his plans to develop the Naze. Due to the absurdity of the system, which protects built infrastructure but ordinarily sees little to no value in agricultural or amenity land, it is likely that Evans' development would have been protected with hard sea defences. On the other hand this would have meant actively destroying the SSSI. One perhaps shouldn't speculate on past events but the result in this case would have been extremely important in terms of planning law. The Nature Conservancy Council would have been forced to protect the SSSI status whereas other government departments may have preferred the decision to place hard defences at the Naze to protect the bungalows on top of the cliff. The victor in this scenario would have been able to change the face of planning law in similar situations across the country.

W.H. Dalton was one of the first to consider seriously the erosion at Walton. In a paper titled 'Walton and Frinton in 1902' he considered factors such as the relative cohesion of the London Clay and overlying Red Crag, the percolation of water, effect of the sea, alternation of dry and wet weather and chemical change in the London Clay. His solution was to construct an esplanade and allow the cliff to reach a slope compatible with permanent stability. He concluded by writing that 'the only effectual system under the conditions is that of closely set groynes rising above springtide level to accumulate soil and check the removal of slipped or solid clay'.[262] However, at a time when the Naze was outside the spotlight, nor protection considered a public necessity, Dalton's remarks meant little.

A further problem that the Naze faces, especially on the northern beaches, is the loss of sand to the sea. Northerly winds that originate from onshore create a strong undertow, pulling the sand out to sea in the winter. The situation is not as simple as this and there have been observations of accretion of material on Stone Point and Pye Sand. However, without a comprehensive survey it is impossible to fully understand the coastal processes involved at the site. During the course of researching for this book many have reminded me of the breakwater system that existed at the Naze for many years, which retained much more sand along the frontage than there is at the time of writing. Similarly I am reminded of the Victorian drainage system which has fallen into disrepair.

The industrial revolution paved the path for the phenomenon of anthropogenic accelerated climate change and this is influencing the volatility of the sea. This will become a far more important issue in the next fifty to a hundred years as climate alters and governments face difficult decisions about which areas to 'save' and which to leave the sea to 'reclaim'. Increasing amounts of carbon dioxide emissions from heavy industry, aircraft and motor vehicles as well as other emissions, particularly from agriculture, are resulting in the melting of the polar ice caps, accelerating sea level rise. Our greed for cheap energy does not seem to be waning and so we can only assume that the problems will worsen over the next twenty to fifty years.

Global sea levels are predicted to rise substantially, reflecting past recorded figures. Over the period 1901 to 2010, global mean sea level rose by nineteen centimetres and the rate of sea-level rise since the middle of the 19th century has been larger than the mean rate during the previous two millennia. Looking ahead to 2100, projections for global sea level rise are placed at anything between twenty eight and ninety eight centimetres, with regional fluctuation.[263] This is due not only to the melting of glaciers and sea ice but also to oceanic thermal expansion in volume due to increased heat induction. The increase in surges and bigger waves, plus the increasing and deepening barometric lows that accompany climate change will further escalate erosion, increasing the risk of coastal flooding. Landowners have been responsible for maintaining their own sea walls for centuries and although some aid from central government has been available at times over the past century, many areas of sea wall are inadequate for projected future scenarios. In some instances, the imposition of physical sea defences themselves, whilst protecting that section of the coastline intended, can aggravate erosion along sections of the shoreline without such protection, such as at the Naze, causing sand and gravel starvation to create hotspots of erosion.

The Naze is made up of soft rocks and is currently protected by the Crag Walk to the east (apart from a large section between the Crag Walk and saline lagoon on the northern edge of the Naze) and concrete sea defences to the North and West. The Tamarisk wall was breached in 1953 and since then the rate of erosion and pressure has greatly increased. The cliffs are eroding at a rate of one to three metres a year. One of the most volatile areas of the wall is to be found on the north east corner. The sea wall was particularly badly affected during the winters of 2013/14 and 2014/15 and at the time of writing the wall is only a few metres from breaching. Since 1953, two hectares of land has been lost to the sea in this area, easily visible when comparing past and current maps. The old Tamarisk wall can be seen clearly out on the beach now, the sea having fully 'reclaimed' the area in front of the present sea wall, a landscape of sand dunes and mud flats. A breach on the north east corner would result in the flooding of a large area of the Naze headland and would mark the beginning of a wider retreat in the Hamford Water area, the Naze currently reducing the energy of the North Sea and preventing it from consuming Hamford Water. Walton Hall marshes, lying within the sea wall, covers an area of very low lying land and makes up about a third of the peninsula's land area in total.[264] Taking the information above, this kind of environment is at a critical level of risk in the new predicted climate. To the north of the sea wall lies Stone Marsh and Stone Point, currently part of the Walton Hall estate. These areas of land, along with the eastern parts of Horsey Island and to the north, Foulton Hall farm, form the three hard points, protecting Hamford Water from the North Sea. Horsey Island was given between 0 and 10 years until the eastern side breaches, in the controversial Shoreline Management Plan of 2010.

Managing coastal defence has been a problem in the Sokens for generations. The Anglo Saxon Chronicle notes an account of a disaster on a national scale in November 1099 when '...on the Festival of St Martin, the sea flood sprung up to such a height, and did so much harm, as no man remembered that it ever did happen before.'[265] J.M. Stratton has further noted a number of 'great sea floods' from the eleventh century, namely on 28th September 1014 as well as in 1015, 1092 and 1100.[266] St Paul's Cathedral also suffered in defending their estate. On one night in the twelfth century they lost several hundred acres of land to flooding. The reference to the disaster is continued in the Cathedral Domesday of 1222, noting that '...at Walton the rents of certain manorial holdings had decayed because the sea had wasted them.'[267] This loss shows that the land to the north east of the Naze, which has been lost since the medieval period was very low lying, enough so that a single breach could reclaim much land to the sea. In 1292 Canon Fulk Lovel's sea wall, built to protect the sheep marsh, was breached and instead he was advised by local people to concentrate his efforts on strengthening the Strode causeway to safeguard grazing on the saltings.[268]

Although the enclosed marshes at Walton Hall were not embanked until much later there is documentary evidence that parts of the Essex marshland were embanked by the end of the twelfth century and there were customs and law in place to protect them. In 1201 for example William Torell brought before the king's justices a complaint against the abbot of Stratford, 'touching a certain dyke (fossatum) thrown down in Little Thurrock'. The word dyke implies a bank with a ditch

131

behind – similar to the 'borrow dykes' behind Essex sea walls today. The defendant pleaded 'that a certain law was made in the time of King Henry the father (Henry II) concerning the marshes, and he prays that it may be observed'.[269] The justices then commanded the sheriff to look in to the 'law of the marsh' and therefore we can tell that statutory protection was in place to some degree in certain places.

Through the thirteenth century local customs regarding the nature of sea defence management came, in many places, more and more under the jurisdiction of royally appointed officials. By 1210 the 'law of the marsh' in Essex embodied the principle that each man should contribute to the upkeep of defences from which he benefited, in proportion to his land or rights on the marsh.[270] Commissions were issued by the king in Chancery as Letters Patent under the Great Seal, and took their authority from the royal prerogative. The first known commission in Essex was to Stratford and it should be said that most commissions were concerned with the coast on the Thames. The first commission for Tendring was not declared until 1451 and even then it is unlikely that the survey covered the marshes at the Naze. It is unlikely because the role of the commission was not to undertake work but to declare whose responsibility it was to undertake the work. It was clear for all that Aedulvesnase was under the jurisdiction of the Dean and Chapter of St Paul's and it would be the Cathedral's responsibility to maintain coastal protection works.

It was in the fifteenth century that the biggest losses of arable land occurred. According to the lease of 1468 'Homefield' was the last remaining of the domainal fields, closest to the manorial complex. Middlefield had been reduced from 218 ½ acres to just 92 acres and Eastfield had been lost completely to the sea.[271] There are numerous references to 'extreme' weather events including heavy rains, high winds and hard winters, all of which would have been coupled with frequent coastal surges. According to Gottschalk, the number of storm surges to affect the southern North Sea between 1400 and 1449 was identified as nine.[272] This compares with seven for 1350-1399, four each for 1250-1299 & 1300-1349 and three in the period 1200-1249. The lands in the Netherlands were particularly badly affected with tens of thousands drowning in the 'Elisabeth floods' of 1404, 1421 and 1424.

It was in the sixteenth century that all parts of the realm were brought under statutory protection in terms of a legal foundation for managing sea defence. An Act in 1515 set out that the 1427 Act, which offered parliamentary authority to the issue of commissions, would be continued 'for ever'.[273] However, in reality local custom prevailed and marshland continued to be engulfed by the sea on many occasions. An increasing number of coastal manors were reclaiming land and building clay sea walls. By the seventeenth century a hotchpotch of different obligations existed, governed by covenant, custom, tenure, frontage and ownership.[274] The commissions of sewers in Tendring governed the recognised 'Tendring Level', which only included the stretch of coastline from Little Holland to Frinton. Walton and the vast majority of the Tendring coast lay outside the commission's jurisdiction. Protection of the marshland in these other areas depended entirely on the attention and means of the landowner or tenant. For example, in September 1691 Sir Harry Campbell of Birch Hall contracted a carpenter and a labourer from Kirby to 'make up a certaine breach in Horsey Isleland lyeing against Saltfleet'. Sir Harry agreed to pay £30 and provide 'timber, broom, boat and boards'.[275] Earlier in the seventeenth century Walton Hall marshes were drained under the supervision of the Dutch engineer Nicholas van Cropenbourgh. Land reclamation projects had been particularly fashionable since the reign of James I who acted as patron for many such

133

programmes and Dutch drainage enterprises were much admired.

The eighteenth century agricultural reformer Arthur Young was a great supporter of using reclaimed land for arable cultivation. He thought the soil at reclaimed Foulness Island to be of the richest in the county.[276] Indeed, once marshland was recognised as being suitable for cultivation it seemed a better investment to keep spending money on draining and sea defence became more important. Towards the end of the eighteenth century a number of developments took place across Essex extending the success of drainage operations. Hollow and under-draining became popular, inserting drainage pipes twenty or thirty inches below the surface, instead of relying solely on field ditches. In the nineteenth century cylindrical tile pipes were invented, advancing drainage technology even further. Nonetheless, Young, writing in 1807, recognised the problems that drainage engineers faced, especially in areas where there were no commissioners of sewers to enforce care, such as at the Naze and elsewhere in Tendring:

> The sea gains on all the coast of Tendring hundred; and it is the opinion of many very able men that the expensive embankments which at present are the security of a large extent of marshes, will be carried away: thirteen years ago, they broke at St Osyth, and the mischief done was very great. Mr Bawtree thinks that they have not yet recovered it. Holland Haven is under the control of commissioners, but they meet with great difficulties in their operations.

The old village of Walton lies somewhere to the east of the present settlement, under the sea. Philip Morant wrote in his *History of Essex* (1768) that local fishermen believed that Old Walton was located out on West Rocks, as they could see the remnants of buildings at low tide.[277] This would place it four miles out to sea meaning that at least this amount has been lost in the past seven hundred years. This is possible although it is likely that old Walton covered a large area to the east and north of the current town. Dr Rosamund Faith and her team of underwater archaeologists from the University of Oxford have tried to examine the sea bed in recent years to solve the mystery of the lost town but failed in their efforts due to poor visibility.[278]

During the medieval and early modern periods saltings, which went far out to the north and east of the current Naze landmass, were submerged. The Chapman and Andre map of Essex of 1777 was the first time that sea walls were presented separately from saltings; those which now lie presumably outside of the walls. Richard Rigby had a grand plan to construct a whole set of coastal defences to protect his Hamford Water estates but lack of funds prevented him from doing so. This was at about the same time he tried to follow through with his failed port and spa ambitions at Mistley. Through the nineteenth century there are various accounts of unusually high tides followed by breaches. A high tide on 17th December, 1846, overflowed a wide extent of land, and is said to have destroyed about £2000 worth of property and land on the estate that at that time was leased by J.W. Hicks. It is written that 'the Terrace wall was washed down in the summer of 1839, and a new one was built in 1840'.[279] In 1856, the sea wall between Frinton and Little Holland was badly damaged by 'unusual violence of the sea' and the land flooded.[280] The diarist and Tendring coastal defence commissioner Richard Stone noted regular surges in the early nineteenth century.[281]

The most significant flood of the 19th Century occurred on 29th November 1897 and was known as 'Black Monday'. The sea wall was toppled in over thirty different places in Walton, completely overwhelming the Naze. The *Essex Naturalist* noted that soon after noon at Walton:

> ...the tide was of the proportion of a spring tide, but it had yet a long time to flow, and there was a strong wind which tossed the backwaters into a very rough condition, forcing the waves against the walls, and in a very short time making breaches, through which the waters rushed, covering the pastures.[282] The basements of houses were soon flooded and traffic on the main road stopped. At the wall around Walton Hall marsh the sea did great damage, breaking through in quite thirty different places, some of the breaches being several yards in length.[283]

Following the disaster the writer of this article toured the Essex coast (once in the spring and once in the summer of 1898) to assess the situation. Many places were in a disastrous state and he noted that several farms between Dovercourt and Great Oakley had been abandoned. Peewit Island in Hamford Water was derelict and the 'luxuriant pasture' of Bramble Island had been turned in to 'an enormous waste of black rotting mud and weeds'.[284] It was soon after this that the site became an explosives works. Further south, Northey Island and New England Island, off Foulness, had been abandoned. The one exception to his assessment however was Walton where he notes that the Naze would long ago have become an island 'but for the constant exertions of the Waltonians, who have spent large sums of money in keeping up strong defences and stone walls'.[285]

There were further severe storms and breaches in 1901, 1906, 1921 and in 1928 when fourteen people drowned in Westminster, Hammersmith and Fulham cellars, awakening the government directly to the issue. It was clear that Essex faced a real problem in the form of coastal flooding. The result of the 1928 floods was the Bledisloe Commission and subsequently the 1930 Land Drainage Act which cleared previous Acts in favour of drainage on a broader basis. However, agriculture, along with other industries, faced a substantial financial stress in the 1930s, as a result of the Depression triggered by the global financial crisis of 1929. Justifying using ratepayers' money to maintain walls which protected low value land and a struggling agricultural industry was difficult. However, clearly the landowners could not pay without putting themselves out of business. The Drainage Commission faced a critical political issue. Could they really justify using ratepayers' money to prop up the physical security of coastal farmers? The vice-chairman wrote openly about the issue, accepting that the general county ratepayer:

> has not any clearly defined material interest in protecting such land. Money spent on making this land safe from seawater will amount to a present bestowed by the general body of ratepayers upon whoever may be the owner of the land in five, ten or twenty years time, for it seems unlikely that he could ever be rated proportionately to the cost of protecting the land...if the sea walls are to be kept up, someone has got to pay. The present owners of the land cannot pay, whether by drainage rate or otherwise; the county ratepayer ought not to be asked to pay for something which is, to a great extent, no real part of his general drainage problem and would be nothing less than an endowment to landowners, some of whom have been able to buy their land for a few shillings an acre simply on account of its insecurity...What is the answer?[286]

Clearly, the Catchment Board faced a difficult decision and raised their concerns with the Ministry of Agriculture. The Treasury made it clear that they were unwilling to grant funds for the scheme. The Catchment Board faced years of having to fund works with often unreturnable loans as well as criticism when walls failed.

Following further severe floods in 1938, 1944 and 1947, the Essex River Board was established in 1951 whose job it was to administer an area upwards of a million acres including 400 miles of coastline. However, there wasn't any way of being prepared for what the county faced in January 1953; a disaster that arguably superseded all Essex floods that had ever taken place before. On 31st January 1953, reports started coming in from all along the east coast that the walls were being overwhelmed long before the predicted time for high tide. The amount of water involved was far higher than the predicted amount and the wind blew at hurricane force. At 8pm in Harwich a gust was recorded as being 81 mph.[287] The entire Hamford Water area quickly became littered with explosives and chemical products being washed from the chemical works on Bramble Island. Horsey Island was completely submerged by 12.30 am on 1st February and was littered with all sorts of chemical debris that had floated across from Bramble.

At 10pm on the night of the January disaster, J.W. Eagle noticed that the sea was almost breaching the sea wall and so he moved his cattle off the marshes to higher ground. The Naze soon became covered, with water gushing down into the town, which was becoming inundated with water entering on three sides, from the Esplanade, the Backwaters and the Naze. The police were out warning as many residents as possible in the danger area of the Walton Gap. They reported to Clacton at 12.30am: 'Kirby Road, Churchfield Road, Saville Street, Standley Road, Mill Lane, all flooded. All persons warned re

138

danger to life. Everything under control.'[288] Boats travelled up and down Kirby Road and the roads around the Walton Gap, rescuing people from upstairs windows. From 1.45am Walton coastguard noted that the tide was receding. Although nobody lost their life at Walton, unlike elsewhere on the Essex Coast and in the Netherlands, 194 houses were flooded to a depth of six feet and a large amount of farmland, not just on the Naze but around Hamford Water, was ruined through salinity. In the wider Essex agricultural community, of the 41,760 acres of farmland that had been flooded, 7,000 had been of winter corn which was afterward not worth harvesting.[289] Pulses and clovers had been instantly killed and most horticultural crops had died, including many hundreds of acres of established orchards. Of the 15,075 acres of flooded Essex arable land, just 2 per cent produced a crop in the summer of 1953, lower than any other county.[290] Across Essex there had been more than eight hundred separate breaches, totalling sixty seven miles of failed walls.

At the Naze, the calamity led to the complete loss to the sea of two hectares of land to the north of the current wall. One can still see the old Tamarisk wall out on the beach. J.W. Eagle noted in his farm diary on Friday 6th February:

> Spent morning with John Fenn surveying walls. Walked the whole length of all walls. Walton Hall wall is very bad indeed with one breach of 250 yards and another almost continuous for half a mile. We decided to get all cattle away if possible.[291]

After the disaster the walls to the west and north were heightened and the concrete strengthened, although this was a long process and the drainage of water from the backwaters and the Hall marshes was very slow. Following through into the modern day, a vast amount of money and time has gone in to strengthening and maintaining the existing concrete walls

139

to hold the line. However, today it is difficult to see a future for this specific aspect of defence. There have been numerous exceptionally high tides since 1953 and the trend pushes only one way; that is towards more severe storms and more frequent high spring tides. The current system of sea defence has to be reviewed to fit in to the changing coastal environment. Crucially, Walton would probably not be able to cope in another surge situation. However, we should be wary of suggesting that the sea should remain in a rigidly defined space at all. The sea is only seen to have 'its place' due to human decision, for various anthropocentric reasons, as to where this should and should not be. The way forward, whatever that may be, should work with the sea, not against it and crucially the plan should be sustainable.

In recent years the eastern side of the peninsula in particular has been under the watch of the Naze Protection Society (formerly 'Save the Naze').[292] For more than three decades, volunteers have worked to raise funds for the 'protection' of the Naze cliffs. In 1975 the 'Save the Naze Action Committee' was formed to 'promote the case for protecting the Naze during the period of local government reorganization'.[293] Notable amongst the members of this first committee were Dora Cohen, Ernest Meadowcroft and Reg Bloom. In 1981 Save the Naze evolved in to the 'Naze Protection Society', holding its first public meeting on 21st January 1981 at the Martello Caravan Park under the chairmanship of Mr E.F (Pat) Kingsley OBE. At that meeting the committee was elected comprising Pat Kingsley OBE, John Barker (Treasurer), Diana Chate (Secretary), Noel Clifton-Brown, David Hall, Mike Todd and Bill Speanes OBE. Sir Julian Risdale CBE MP was the first President of the Society and Mr Amedee Turner QC MEP the Patron. A booklet was produced soon after, funded by a number of advertisers and printed at cost by Flora Hunter and the staff at the Walton and District Advertiser, to act as a guide to the flora, fauna and bird life to be found at the Naze and to raise awareness of the NPS. In the booklet an article written by N. Clifton-Brown suggested the demands of the Naze Protection Society at the time. They were firstly calling for 'the construction of a concrete or stone shoe around the north east corner and for the extension of a breakwater.' There would also be 'a programme of manual draining and subsequent deepening of the channels so that water could be led off the cliff into the sea'. The cliff would then be planted with 'grasses and shrubs to anchor the soil and to give further drainage to the deeper regions. The varieties of plants considered suitable for this exposed situation include Grey Alder, Sea Buckthorn, Goat Willow and Tamarisk'. The NPS called for a 'sponsor a tree' scheme to finance this project. The article also called for experimentation in the use of 'Reno mattresses – flat wire baskets filled with stones) and the use of groynes'.[294] None of these calls for action were successful however and it wasn't until thirty years after the formation of the society that words were converted in to action through a partnership project resulting in the 'Crag Walk'. Despite this, in the thirty five years before the construction of Crag Walk many attempts at finding solutions were made.

In 1977 the Department for the Environment established an inquiry drawing up a plan for a £1.4 million project to safeguard the Naze cliffs.[295] However, this never came about due to the claim that 'lives and property were not at risk'.[296] The project would have seen the construction of 1020 metres of precast concrete blockwork revetment along the foot of the cliff, eighteen timber groynes and one permeable groyne as well as artificial replenishment of the beach and subsoil drainage works both at the cliff top and on the cliff face. In short, the Naze cliffs, which by then had been designated its SSSI status, would have become an extension of the concrete protection of the rest of the urban district. On 22nd November

1978 a public inquiry was held, at the request of the Department for the Environment, directed by Mr W.B Kirby, to consider the merits of the scheme. The Inspectors conclusions were as follows:

The scheme, amended as suggested, would meet to a large extent the conflicting interests of the principle parties and if the goodwill shown at the inquiry were continued there is no reason to suppose that a wholly satisfactory outcome could not be achieved.

The Council's evidence on the causes of erosion are of little misleading, but I and my two assessors are satisfied that the cliff is being eroded by the sea although we believe that the rate of erosion is being accelerated by the seepage of surface water into the gravels and Red Crag causing slips in the cliff. The purpose of the concrete revetment and the groynes is to prevent erosion of the cliff by the sea, whilst the purpose of the cliff drainage is to stabilise the cliff. If the concrete revetment were constructed without stabilisation of the cliff the revetment would be in danger from future slipping of the cliff, therefore the stabilisation of the cliff is required if the revetment is constructed and it is in my opinion coast protection work.

The Nature Conservancy Council have asked that the scheme be postponed for two years but in my view time is of the essence and if the scheme is approved then it should be carried out as early as possible. However, there will be some delay before the decision is reached, time necessary for the preparation of contract documents, approval of the successful tender and other administrative matters and this interval together with the Council's willingness to commence work on the northern end should be sufficient to meet the requirements of the NCC. The proposed 'bench' would help to stabilise the cliffs and ought to be included.

Turning now to the need for the scheme I am mindful that hitherto available financial resources for coast protection works have been directed towards the protection of urban areas and the cost of any proposed works has to be weighed against the values of the land and property being protected. If this criteria is applied then no case has been made out and I have no alternative other than to recommend that the Secretary of State direct the Council not to carry out the works proposed.

Nevertheless I was very impressed by the sincerity of those trying to Save the Naze and by the widespread support for this project. The Naze is an area of high amenity value and one must ask whether such assets, which cannot be measured in monetary terms, should be allowed to be destroyed by the sea. I propose to recommend that the Secretary of State should examine the merits of this case to see whether exceptionally it could be approved. If the scheme as amended cannot be approved then the Council should be invited to submit a scheme to protect the northern flank of the Mabel Greville works and to consult with the Anglian Water Authority with a view to preparing a joint scheme designed to protect the end of the Authority's flood prevention wall.

The Secretary of State eventually decided that the available financial resources for coast protection works must continue to be directed towards built up areas and that it was not possible to make an exception in the case of the Naze. The Council were therefore instructed to go no further with the proposed works.

Further plans were drawn up in 1979 and 1985 but both ultimately failed to come to fruition. The 1985 plan was particularly well received by the Department of the Environment as it addressed most of the concerns that had been raised in 1977 but the key issues were based on a disagreement as to how funding should be divided between central and local government and objections from the Nature Conservancy Council on the grounds that fossils in the cliff would no longer be exposed.[297]

Pressure remained however to take some sort of action against the erosion of the cliffs at the Naze. In August 1986 Tendring District Council wrote to the Department of the Environment to notify them that they were having meetings with the Nature Conservancy Council about their objections. Only by having this dialogue would a compromise be achieved. At a meeting of the 'Control and Protection Committee' in October 1986 it was agreed that Strathclyde University should be commissioned to carry out a feasibility study for the drainage of the cliff top. This was set in motion and a report was published in 1988. However, once again no direct action came about as a result.

In the mid-1990s further attempts were made to act at the Naze. On 7th October 1996 a public meeting was held at which it was agreed to set up a 'Naze Trust' to seek funding opportunities. On the 25th October a follow up meeting was held in which it was said that since the public meeting there had been 'numerous correspondence seeking support from various bodies such as the Countryside Commission, Harwich Haven, Essex County Council, the Protection of the Countryside for Rural England and the MEP Anne McIntosh'.[298] Terry Norman, who chaired the meeting, explained a potential action which involved the construction of three hard rock points – one at the end of the sea defence by

144

the Tower, another at the beginning of the sea wall near the sewage works and another in between these two points. It was suggested that this should be funded by the Environment Agency, MAFF and £150,000 from Tendring District Council who had already set aside this amount for work at the Naze. District Councillor Peter Lawes was asked to follow up this suggestion. At the 25th October meeting the 'Naze Residents Association Committee' was enlarged so that committee members included Terry Norman (Chair), Terry Lawrence (Treasurer), Sharron Moulson (Secretary), Tom Walters (Publicity), Ken Coleman, John Titchmarsh, David Eagle, Ray Smith, Joan Bunting, Gladys Furminger and Ron Wyatt. The main aim of the committee was 'to seek funding for the implementation of the three hard rock point scheme'. The committee also wanted to encourage the Naze car park to be run by volunteers with all proceeds going towards 'saving the Naze'.

This was the case until November 1998 when three hundred tonnes of Leicester granite was placed on the north side of the Tower Groyne at the Naze as part of the first stage to protect the Tower.[299] However, critically the second stage of this particular plan was never followed up. In 1999 when Harwich Harbour was dredged a substantial amount of sand was deposited in front of the cliffs on the north and north east of the peninsula as well as at Stone Point. However, without the breakwaters to keep it in place it did not remain.

In 2001, Tendring District Council awaited results of a costly study in to the viability of 'fish tail' hard points to be placed at strategic intervals at the base of the cliff. It was hoped that this approach, which had proven successful further down the coast at Jaywick beach, would be viable for the Naze. However, the report concluded that the Naze is too exposed to easterly winds for the scheme to be workable.[300] In 2003 it was proposed to extend the Tower groyne revetment but, yet again, funding became an issue and the money did not come available. In 2005 David Gager, the Chair of the Naze Protection Society, approached the Council to bid for lottery funding for a new project, taking a radically different approach to tackling the problem. The project entailed building a 'road' in front of a 200 metre length of the cliffs, providing an accessible way for members of the public to view the cliffs. This project took on a number of other partners and evolved in to the 'Crag Walk', which was eventually completed in 2011. The Naze Protection Society, a registered charity, worked with partners including Tendring District Council, the Essex Wildlife Trust, Natural England and the owners of the Naze Tower and Walton Hall farm to construct the one hundred metre retaining wall. It is described by the Naze Protection Society as an 'educational public access viewing platform'.[301] The structure allows the sea to encroach it and percolate around it, whilst minimising the effects of erosion.[302] The partnership believe that the scheme will protect the Naze Tower for another century although erosion of the cliff will continue further along the stretch of coastline up to and including the north east corner, threatening the sewage works which lies within the flood plain.

When writing on the Naze it is sometimes easy to take it out of geographical context but the sea is vital to its story and has, over the years, presented as much of a risk to local people as it has to the headland itself. In 1884 (to quote the minute book), 'a meeting of gentlemen was held at the Vicarage, Walton-on-the-Naze on Friday 12th February, where appointments in connection with the above named branch of the R.N. Lifeboat Institution were determined upon'.[303] The members of the local branch committee were Sir John Henry Johnson (President), Mr Robert Warner (Chairman), Mr J.W Eagle (Hon. Secretary), Mr Davie, Mr Dorling, Dr Fowke, Rev. M.S Horton,

Mr Walter Howe, Mr Benjamin Latchford, Mr Randall, Mr Rickarby, Mr Sheppard, Mr Wilmshurst, Rev. J.T Cooke and Mr L.B Trimen.[304] Appeals were launched in the town to provide subscriptions to pay for the cost of the boathouse, slipway and annual running costs. Construction of the boathouse began in the Autumn of 1884, situated next to the Coastguard Station at the end of the East Terrace.

Later that year the first R.N.L.I boat arrived at Walton – a thirty seven footer Pulling and Sailing self-righter built at Limehouse by Forrest and Son and paid for by the Dramatic Club of the Honourable Artillery Company – she was named the H.A.C. She was launched and christened on Tuesday 18th November in a ceremony led by the Baroness Bolsover, mother of the Duke of Portland. The first crew were Henry Britton (Coxswain), Robert Terris (2nd Coxswain), John Terris (Bowman), Haven Puxley, Abraham Halls, William Hammond, Jeremiah Bates, David Polley, George Hatcher, George Bates, John Sharman, Charles Bates, John Hatcher, Henry Cutter, Fred Batchelor, William Popperwell, George Cooper, William Oxley, John Fairbrother, A. Barnden, George Clamp, John Downes, T Frosdick, W Sparrow and W Norton.[305]

Unfortunately on the H.A.C's first call out the crew were unable to find the source of the distress and she was also damaged against the bottom of the slipway. The event was later reported in the local Gazette:

> After the return from the nocturnal journey on Wednesday, the 19th, the Lifeboat was seriously injured. It appears that the boat was being brought in at about midday, when, owing to the defective arrangement of the slipway already referred to, it was found almost a matter of impossibility to run the boat direct upon the slipway. Two or more attempts were made. On the last attempt, as the nose of the boat

touched, a heavy sea struck on the her beam on the stern end, bowling her round broadside on the slip, at the same time forcing her bow on to one of the butt ends of the curb of the slipway, making a hole on the port side of her bow some sixteen inches in diameter, crushing through the outer covering and the inner skin. The Inspector of the National Lifeboat Institution was soon on the scene and directions were given for the repair of the boat. It is understood that the damage may cost £40 or £50. No doubt the slipway will be properly seen to, so as to avoid a similar occurrence in the future. It is not near wide enough, and nothing like being long enough. The top end is exceedingly dangerous at night or any time, and there is not sufficient room to lock the carriage or for the people to assist in getting it down. It certainly is remarkable that someone who assisted in getting the boat down the night previous did not sustain injuries or that no one fell over the breastwork. The matter should be at once seen to, for instead of taking only about ten or fifteen minutes to launch the boat, it took almost one hour.[306]

Then, on Boxing Day 1884 the crew made one of the Lifeboat's most memorable rescues when she took twenty five men from the German ship Deike Rickmers, which had run aground on the Long Sand. The rescue took over twenty four hours in total, the Coxswain forced at one point to drop anchor and wait for four hours in poor weather conditions, before there was sufficient water to cross the sands off the Black Deep. The Gazette reported afterwards that 'the reception, which the boat met with at Walton, is almost indescribable. It is sufficient to say that the scene was one of the utmost excitement, emotion and thankfulness.'[307] The H.A.C ended her service after sixteen years and was replaced in 1900 by the

James Stevens No.14, which today, following restoration, is recognised as the oldest surviving motor lifeboat in the world.

The James Stevens served for twenty eight years as the Walton and Frinton lifeboat and during that period saved two hundred and twenty seven lives, in a hundred and twenty six launches. James Stevens was a Birmingham businessman, who left a legacy of £50,000 to the RNLI when he died in 1894. He wished the money to provide twenty lifeboats, all named after him and number 1-20. In 1928 came the E.M.E.D which served through the war. Because of the fear of invading forces part of the pier was destroyed and the lifeboat was moved to the Backwaters. Further, due to it being impossible to use maroons to summon the crew, foot messengers were used to summon the men. In 1953 the E.M.E.D was replaced by the Edian Courtauld, a gift from Augustine Courtald, a member of the RNLI's committee of management, in memory of his mother.

This boat was a forty six foot nine inch by twelve foot nine inch Watson type cabin boat built at Whites of Cowes on the Isle of Wight. This period, up to the early 1970s was the busiest in the history of the Walton lifeboat. Indeed, in 1966 the Walton crew were the busiest in Great Britain with thirty three call outs.[308] Most of these calls were due to the pirate radio ships off the coast of Frinton. For example, on January 19th 1966 Radio Caroline broke free from her moorings and blew ashore near Great Holland. The crew, led by coxswain Frank Bloom managed to board the vessel in terrible weather conditions. The District Inspector, Lieutenant Commander D.B. Cairns, RNR, stated in his report that 'the coxswain and his crew showed courage, determination and skill in boarding the lifeboat in conditions of wind, sea and bitter cold which were the worst known for many years at this most exposed station'.[309]

149

From 1977-1983 the Earl and Countess Howe was stationed at Walton, previously the lifeboat of Great Yarmouth. From 1984-1993 the City of Birmingham patrolled the waters, having been previously at Exmouth. Then from 1993-1996 the Sam and Joan Woods and from 1996 the Kenneth Thelwall II has served. This book is not primarily about the lifeboats and so will not go in to any further detail. However, anyone interested should read a copy of John Steer's *Walton and Frinton Lifeboat: A Station History 1884*-2005 which sets out in detail the history of the lifeboats, touched upon above.

A number of medals have been awarded to crew members of the various lifeboats through the years, including Bronze medals to John Byford (1917), Walter Jonas Oxley (1939), Fraser Bacon (1939), Frederick Williams (1939), Thomas Bloom (1941), Thomas Brooke (1941), Frank Bloom (1966), Keith Richardson (1973) and Jack Barrett (1973). 2nd Service Clasp Bronze to Thomas Bloom (1945) and Walter Jonas Oxley (1964) and Silver medals to Henry Britton, the first coxswain of the Walton and Frinton Lifeboat (1902), William Hammond (1917), Thomas Bloom (1939) and Frank Bloom (1975). The Lifeboats have also been ceaselessly supported by a voluntary team of support and honorary workers, many of whom have also been recognised by the RNLI. John Eagle was made an Honorary Life Governor of the RNLI in 2005, having served as Chairman and President of the local branch. His father before him had been made an Honorary Life Governor in 1968. A number of gold badges have been awarded – to Brian Rowlen (2005), Philip Oxley (2003), Barbara Pearce (2003), Annette Rayner (2003), Beryl Robertson (2003), Muriel Wilberforce (1974), Dr E Clifford Johnson (1971), Mrs EC Brooke (1964) and silver badges to Susan Berry (2004), Jean Halls (2003), Valerie Bloom (1997), Mrs EA Coe (1977), Ronald Brayley (1976), Miss B Boustred (1974), Mrs DE Davis (1971), Mrs KR Brayley (1964), Mr EC Jones (1963) and Victor Scammell

(1960). The names above are included as testimony to the immense service that these people have provided for their local community and for those in need of help at sea. The activities of the lifeboat crew continue to be difficult and dangerous. Their determination to save lives in the face of these difficult circumstances and the continuation of this great national tradition and institution is vital for a coastal community and it remains a vital service that Walton provides for the local area.

Seven nautical miles off the coast of Walton is situated the Roughs Tower (formerly HM Fort Roughs), one of four naval forts designed by Guy Maunsell to protect the Thames Estuary during the Second World War.[310] These forts form a key, if sometimes overlooked, part of Essex maritime, social and military history and so it is important to outline this here. The forts were built at Red Lion Wharf in Gravesend in 1940 and sunk off the coast of Harwich and Walton on 11th February 1942. The forts consist of two reinforced concrete towers, eighteen metres in height and seven metres in diameter, topped with a steel platform. Each tower was divided in to seven floors of which four were originally used for crew quarters. Up to a hundred and twenty men were deployed to the Towers and the aim was to watch for enemy submarines and aircraft. Those deployed would spend a month on the Towers and then a month on the mainland. The Navy abandoned the towers in 1946 and they stood growing derelict for twenty years. On Christmas Eve 1966 Roy Bates, a former infantry major in the first battalion Royal Fusiliers, raided the tower and took control of it. On 2nd September 1967 Bates declared independence as the Principality of Sealand, a sovereign state with its own passports, stamps, flag and national anthem. He proclaimed himself the monarch, Prince Roy and his wife Princess Joan. The following year Prince Roy's son, Michael Bates fired a warning shot at a British Trinity House vessel which approached the towers. This led to

Roy Bates being arrested when he next arrived on the British mainland. However, the case was dropped when the courts ruled that they had no jurisdiction outside British territorial waters. In 1978 the fort was invaded by a group of Dutch and German businessmen who claimed ownership and took Prince Michael prisoner for three days. Roy Bates counterattacked and captured a German prisoner who was held in the on-site Sealand prison whilst negotiations with the German authorities took place. In 1987 the British government extended its territorial waters from three to twelve miles, thus rendering the Principality's legal claim dubious. However, the principality continues its claim as an independent state due to its founding and de facto recognition previous to the 1987 declaration. Nonetheless, no state officially recognises the Principality. In 2007 Sealand was placed on the market at offers in the region of 750 million Euros, a price suggested by a Spanish estate agent. However, a purchaser has (perhaps predictably given the huge cost) not been found.

This chapter began by suggesting that Walton's cultural history cannot be understood without understanding the profound influence that the North Sea has on the terrestrial environment of Walton and the Naze. It closes by reiterating this point. The sea is a natural force and we should acknowledge that natural spaces are dynamic spaces that change over time. Environmental historians recognise this. Andy Flack of Bristol University has said that 'humans manipulate environments and the flora and fauna within for their own ends. We are used to doing this. Consequently, when natural disasters decimate spaces which we have designated as 'ours', we are often profoundly disturbed. We are no longer used to vulnerability in the face of the elemental forces exerted in the rest of the natural world.'[311] This is where I struggle between my capacities as an historian interested in environmental dynamism and in my desire to see a sustainable future for the Naze as a space where human activity can be prolonged. The answer to the question of whether to 'save the Naze' or not depends on one's perspective and background and is rarely a black and white response.

CHAPTER 8

Conclusions – 'Saving' the Naze

At the very northern tip of the Naze lies Stone Point, a uniquely remote spot, albeit still accessible by walkers willing to risk the tide and by yachtsmen who can moor their boats nearby. This sandy spit acts as a physical protection for the Naze and Hamford Water situated behind. It forms a hard point 'protector' that regulates erosion and accretion of material elsewhere on the peninsula. The sand itself has moved and shifted significantly over time and in recent years has sometimes been a cause of concern, with some claiming it will move to close off the Walton Channel and threaten Walton's yachting industry. This shifting is echoed in the Naze's changing cultural identity, which has also evolved from a place of prestige within the Soke to a symbol representing fragility of landscape, set against a 'threatening' sea. It has become an 'endangered' place in need of human protection, based within a town facing its own crisis of identity in a post-modern world.

The idea of 'saving the Naze' is only a recent phenomenon, albeit with roots stretching back to early coastal defence plans, outlined in chapter seven. The drainage operations of Sir John Johnson in the nineteenth century and Canon Fulk Lovel's thirteenth century sea wall, built to protect the sheep marsh (which ultimately failed), are similar to current actions of defence in terms of their aim, which was ultimately to protect land from being 'claimed' by the sea. However, critically, it is attitudes that have changed more substantially. In 1292 St Paul's Cathedral concentrated their efforts on protecting one part of the Naze above another, yet today there seems to be a determination to maintain the line across the peninsula.[312] There appears to be a communal desire to 'protect' the Naze, with multiple parties coming together under a common banner to 'save' the Naze for posterity. Further, in the 21st century, the Naze is valued for different reasons than it was in the thirteenth or eighteenth. The development of professional science imposed a geological value on the Naze site, previously unrecognised. The use of the site as a place of leisure, residential housing and 'built heritage', as land use increasingly moved away from being centred on agriculture, granted other peculiar values not previously envisaged. These different narratives have come together to result in a movement to 'save the Naze'.

The Naze has come to represent Walton and forms a key part of its identity, allowing a seaside town that continues to rely predominately on seasonal tourism, to highlight its 'point of difference' within the wider tourist marketplace. Recognition of this economic importance has helped to drive cultural attitudes, encouraging people to physically 'defend' the Naze 'against' the sea. The proposed developments of the 1950s and 1960s encouraged the public to take ownership of the headland, tangibly in the form of the purchase of the public open space in 1963 but also emotionally and culturally. Local people had been walking, fishing, shooting and poaching on the Naze for many years previous to its purchase by Essex County Council. However, far fewer people visited prior to this and so the Naze was not as much a part of the cultural identity of Walton, as it is today.

By constructing a narrative that the Naze is 'endangered', a myth of peculiarity has been created about the headland, setting it apart from the rest of the landscape, defining its borders and reducing the importance of its links with other places. The Naze is separated from the rest of the coastline in a hierarchy of landscape and has been granted unofficial special status as an 'endangered' asset.[313] Labelling the Naze as 'endangered' has come about as a result of centuries of cultural association, accelerated since the 1950s, with increasing numbers of people having visited the site and having formed their own memories and associations with it. Attachment to place through personal association is a principal means in which a place is considered as having intrinsic value but it also encourages the creation of myths of peculiarity. Places that we know personally act as our own 'safety net landscapes'. In returning to a particular place on a regular basis we build our connection with it and with other people who share similar experiences. Our collective memories are then bound together, through the construction of a shared narrative. To protect this narrative we feel it necessary to physically 'protect' the landscape associated with those experiences – to protect the setting of the narrative. In doing so, the peculiar characteristics that brand a place as being unique are constructed, defined and emphasised and only then can an argument be established to 'protect' it from change.

This book urges people to recognise that the Naze needs to be seen in the context of its landscape and ecology, set within the wider Hamford Water, Thames Estuary, Tendring Hundred, north Essex, East Anglian, British, European, North Sea and global contexts, but also within the context of its cultural, social, maritime, political, agricultural, economic and environmental histories. The Naze is perceived as being pushed to the edge because of the values imposed upon it as well as the environmental change it faces. Evolution of landscape is inevitable. Rather than 'protect' landscapes we must learn to adapt to work with them. Despite suggesting that 'saving the Naze' is based on a myth of status, based on contemporary understandings of the Naze, for many people today the Naze is important and is therefore worth 'saving' for as long as possible. It has become associated with several different and significant categories of 'value'. Nonetheless, we should be wary of placing too much value on one place for the sake of its own peculiarity, when it is intimately connected with other places that tell stories of no lesser or greater importance. The Naze should not be valued in isolation but as part of a broader network of places and narratives.

When government inspector Lovell visited the Naze in 1961 to assess the appeal following the rejection of Godfrey Evans' development plans, he placed the Naze in a 'special category' in terms of its amenity value, scientific value and 'peculiar situation'.[314] Lovell recognised the value of the site in terms of its association on a broader scale. Today, similar categories of value are placed upon the Naze. For example, the Naze Heritage Project aims to encourage people to learn about the peculiar value of the Naze site in terms of its history, wildlife and geology. Particularly in terms of geology the Naze is of great interest. However, its importance should be seen in relation to the geological project as a whole, a modern phenomenon in itself, making up a small part of a wider sphere of knowledge, rather than simply part of the unique character of the Naze itself.

Over the period of research for this book, something between six and ten metres of land has been transferred from landscape to seascape at the Naze, crumbling off the edge, on to the beach and the sea bed. This figure is however a small part of a much longer process of erosion which has been happening for hundreds of years. Despite 'saving the Naze' being a recent cultural phenomenon, the act of defending against surge tides and prospects of flooding have a long history. Behind the coastal defences at the Naze lies centuries of human perseverance, in what has been a perpetual struggle against the results of stormy high tides. Each generation has faced similar problems but it has been since the formation of 'Save the Naze' in 1975 that the pace of the effort to plan for the 'defence' of the Naze has expanded so significantly. The principal emphasis of 'defence planning' has however, so far, been based on protecting the eastern side of the peninsula, particularly regarding how to 'protect' the cliffs. This has recently culminated in the construction of the 'Crag Walk' project, the latest in a long line of projects relating to coastal 'defence' at the Naze. The northern, north eastern and western parts of the Naze are mentioned in the literature and wider debate far less.

Concrete sea walls such as those that run from Foundry Hard in the south west of the Naze to the 'north east corner' are effective protectors of the land behind them but, critically, they require maintenance and also facilitate coastal squeeze, a phenomenon whereby salt marsh, an often under-valued part of our natural coastal defence system, migrates landward and is eroded due to the existence of hard sea walls which act as buffers. At the Naze, like much of the rest of the Essex coastline where land has been reclaimed, salt marsh is prevented from migrating. Salt marshes are effective dissipaters of wave energy and already provide the first line of defence against the tide in Hamford Water, at Stone Marsh and Horsey Island, particularly during stormy conditions. Maintaining the line in the longer term will inevitably lead to the loss of salt marsh in Hamford Water, increasing wave energy and removing a vital natural buffer.

In recent years environmental policy makers have favoured an approach of managed realignment of the coastline to prevent 'coastal squeeze'. In July 2010 part of the sea wall at Devereux Farm, Kirby-le-Soken, which is on the southern coast of Hamford Water, was breached as part of this policy. A number of environmental benefits come about as a result of breaching sea walls, including creating space for new inter-tidal habitats, mudflats and salt marsh. These are valuable habitats for several reasons. Firstly, inter-tidal habitats support specialized plants, often declining in population. They are also important fish nursery areas, spaces for the development of insect and invertebrate populations and, perhaps most significantly, of momentous value for the feeding and breeding of many bird species.[315] Managed realignment reduces the capital and maintenance costs of hard defences elsewhere. The policy used to be called *managed retreat* in Britain. This term has since gone out of favour, as it implies negativity in coastal management. From the schemes that have been completed there appears to be many benefits in carrying out managed realignment schemes. Nonetheless, there is disagreement as to how effective the policy is in creating new saltmarsh habitat.

Walton Hall marshes have been on the front line of the many breaches that have taken place over time. In 1953 many plants and grasses were killed through excessive water cover and salinity and although a breach on that scale has not been suffered since, it is rare for there to be a winter when a large area of the marshes is not flooded. The sluices along the wall increasingly have to be open for much of the winter to keep the land drained. The marshes are important habitat for a number of different birds, invertebrates, small mammals and plants. Ultimately, it must be decided whether to keep this valuable habitat and grazing marsh by strengthening defence systems, or breaching the wall and letting the sea reclaim them. Land management is about assessing risks, establishing a plan of action for both the near and distant future, and then putting that plan in to action. There is currently no such plan. A long term vision for the Naze needs to be created. Cooperation and direct action need to be at the forefront of decision making. I hope that this book will help to trigger a serious debate involving all parties, land managers, government agencies, academics and local people, to work together to establish a coherent plan for the Naze's future, looking at a landscape scale and across the peninsula, rather than focusing on piecemeal aspects.

The Naze has become a symbol for the town of Walton-on-the-Naze and a place of recreation for local people. Even though some Waltonians may visit the Naze more than others, for many it is a defining part of the town that sets the settlement apart from other local towns and villages, giving it a peculiar identity. From the nineteenth century the town has been known as 'Walton-on-the-Naze' rather than 'Walton-le-Soken', suggesting its identity became more rooted in the peculiar importance of the Naze headland, increasingly lacking commonalities with the rest of the 'Soke', Kirby and Thorpe. The Soke itself has died in all but name, recognisable when entering and leaving Kirby-le-Soken and Thorpe-le-Soken. The Naze has different meanings for different people. For me, as I reflect on my own personal experiences, it is a place of great personal meaning and a place where I was able to develop my passion for environmental issues. It is an open playground for children to run, and learn about and appreciate the natural world, for enthusiasts and experts to practise natural history and geology, for local people to walk their dogs, take photographs, to picnic and to act as a centre for community. As populations grow more urban in outlook and increasingly 'disconnected' from the natural world, the Naze provides a space for people in this corner of England, an area famed for its eastenders, its piers and its ports, to appreciate and recognise the natural world on a large scale, as part of their own identity and community. Other than the sea, the geological make up and the sky, the Naze landscape is entirely man-made. Walton's history is heavily influenced by the natural world within which the town is set, yet few would recognise these surroundings as key to the town's identity. It is here however that Walton cannot escape a much wider national and international phenomenon – a growing appreciation and understanding of the natural world, of ecosystem services and how community, ecology and economy interact. Walton's human history has directly influenced and been influenced by the landscape at the Naze and the influence of the North Sea. The Naze landmass is being 'pushed to the edge' physically by erosion and sea wall overtopping but this book recognises that the town of Walton itself is also being pushed. Walton has come to rely on the Naze to define itself and as current land use is threatened it faces its own crisis of identity.

Appendix:

A Series of Oral History Interviews

Oral history is the recording of people's memories, experiences and opinions. It brings an opportunity to gather the views and memories that show an individual's unique perspective of the past and of past events. It can also be useful as a way of intensifying debate and challenging our own views of the past and the future. In 2008 and 2011 All Saints Reading Group, spearheaded by Valerie Bloom, instigated two separate publications of 'memories of Walton', the result of numerous oral history interviews carried out to document memories of past town life. Some of these touched upon life at the Naze but I wanted to go further and centre the story on the Naze part of the town as well as conducting longer interviews in more detail and setting these out verbatim as far as possible within the text. There are seven interviews presented below, representing a cross section of people all with personal connections to the Naze or the surrounding area. The people interviewed either approached me or were approached by me and kindly agreed to outline their opinions and memories. The cross-section of participants is not designed to show a representative sample of people connected with the Naze, for this is a very diverse group of men and women, young and old, local and non-local. Instead, it represents a group of people passionate about the local area, care deeply for its future and are keen to relay their knowledge and memories of it.

Mike Cranstone-Todd, known by many as 'the Nazeman', has a huge amount of experience at the Naze and his life has been directly connected to it in so many ways. During our conversation he told me about his connections with John Weston, with the early days of the Naze Protection Society (Save the Naze), nineteenth century Walton Hall, childhood memories of the Backwaters, his perceptions of Walton and the future. Bryan Hicks spoke to me particularly about wartime Walton, community life, his schooling, the boating lake and the changing nature of the town among other subjects. Mike Pratt, son of Bill Pratt who was the gamekeeper at Walton Hall, lived at Creek Cottages as a boy and was able to relay many stories about what it was like to live and work on the Walton Hall estate in the latter half of the twentieth century. I also spoke to John Halls MBE, a life-long (and well known locally) resident of Walton who has a wealth of memories about the town, to John Fleming who has farmed and lived at Birch Hall, just outside of Kirby-le-Soken for much of his life, suggesting again that the Naze cannot be studied in isolation and to Isobel Ruckert who grew up next to the Samuel Lewis Convalescent Home. I also spoke to my father, David, about his own experiences at Walton Hall and his thoughts for the future.

1. Mike Cranstone-Todd

I spoke to Mike at the Naze café on a sunny but blustery morning in late September. Mike is a naturalist who grew up at the Naze and has dedicated much of his life to studying the Naze and relaying this knowledge to others.

BE: *Can you first tell me about your background, how you first came to know the Naze and relay some of your earliest memories of the place?*

MCT: I'm a fifth generation local, so just about local. My mother's family first came here in 1826. My first memory of the Naze is when I was four or five years old and I came here for a walk with my parents. My interest was sparked when I was eight or nine and I met a funny little man who liked birdwatching. That man was the late great John Weston, perhaps the best naturalist that Essex has ever produced. After that meeting I came up every day. In fact there aren't many days, other than when I lived elsewhere that I haven't been here. I've seen it change from a desolate, barren wasteland in the early sixties after it was bulldozed flat, to the lovely area we have now: secondary woodland and one of the most amazing acid grasslands anywhere in Essex.

BE: *What are your memories of John (Weston) in the Fifties and Sixties?*

MCT: There aren't words to describe Weston I'm afraid. He was just John. Small and manic would be the best way to describe him, with a wicked sense of humour. He was certainly one of the most fantastic naturalists I've ever known. He tried to get us all in to birdwatching but I recognised pretty early on that birds fly away which is why botany became my main subject. John was an amazing character. He enthused you. Small in stature but massive in terms of what he did for the Naze. He was one of the people who fought the development here in the early sixties. There was a consortium headed by Godfrey Evans who wanted to turn it into a holiday complex. There was however great opposition. I think it was probably because locals had been able to access the space from the early fifties. The golf course had been long gone but people decided that they'd like to carry on coming up to the Naze. It was their 'park', for want of a better word, there not being much open land in Walton.

BE: *So in 1963 you didn't notice much change because people were already coming here.*

MCT: Exactly. People were driving all over it in cars. It's hard to imagine now but in 1963 [following the war and several failed planning applications] there wasn't a tree standing anywhere. In fact, the mounds where they pushed most of the debris are still there under the gorse and the trees. Most of them have been turned into rather nice rabbit warrens but they are still there.

I also remember going out to the Tamarisk wall, to sit and eat a sandwich before going home for a late lunch in the afternoon. Because of course the Tamarisk wall was not destroyed in 1953. A small hole was punched in it and then the facing was stripped off and taken to Jaywick, letting it fall apart and putting the new sea wall by the cliff. However, it took a little while for it to go.

BE: *Can you describe your family's experiences of 1953 in Walton?*

MCT: My parents at that time were living with my grandparents in Standley Road, the only house by the school, now called Unicorn House, and I remember my mother saying that when she woke up she looked through the back window and there were three coffins floating by like canoes. They had come out of Oxley's yard, floated across the Mere and ended up on the Bath House Meadow. There were also caravans floating around on the meadow in to the back of the Bath House yard. The clean-up took months. The properties in North Street were particularly badly damaged.

BE: *You've mentioned to me before that your godmother had links with Walton Hall farm. Can you speak a little more about that?*

MCT: My godmother (Miss Barr) lived next door to us in Naze Park Road. She died when I was about fifteen. I remember speaking to her one day and she said how she and her brother Herbert were taken to school in the coach from Walton Hall –

165

perks of the job with dad being coachman! She never forgot how, going past what is now the Naze fish and chip shop, at the bottom of the hill, they'd meet the poor coastguard children walking down from Stone Point. She felt sorry for them as she rode past them in the coach. This would have been around 1880. When you were the child of a humble coastguard you were even lower down the scale than your parents and therefore you wouldn't get a lift to school, you had to walk. When you consider where the school was, now the library in the Highstreet – those kids must have been fit. It always amazes me when you see pictures of Blossom, the last watch vessel that was out there and you count the chimneys. There were six families living there on that boat. When you consider the average Victorian family size and that the boat was marginally bigger than a Thames barge. Around 1900 it was decided that it was unsanitary so they built what are now Creek Cottages (in 1903). However, soon after it was sold to the Hall estate (in 1923). The current coastguard station was built and the coastguards moved with it. The watch vessel was however at Stone Creek for a while longer. Three or four years ago I had an email from a doctor in Margate. His grandfather, William Wrabson, was on the boat when it was taken away and as a result the Wrabson family ended up in Margate.

My godmother died around 1970. She never married and lived at number 17, Naze Park Road and looked after her brother. He'd been on HMS Conquest in WW1 during the Battle of Jutland and was badly injured. He came home, took to his bed and stayed there for the next thirty years. She looked after him until he died and then stayed there until she died herself. The house was named Conquest after the boat. She was a lovely lady even if mad as a hatter!

BE: *I know that you were involved in the early days of 'Save the Naze' - what became the Naze Protection Society. I'm interested to hear what the politics were like, who was involved and how the idea to form it came about in the first place?*

MCT: It began with a public meeting at the Martello to 'save the Naze'. I don't think we were aware of whether it was really possible to 'save' the Naze because, in reality, unless you put expensive drainage in you're never going to 'save' it. We concentrated on erosion. In the early days the main person behind it was John Barker who owned the Marine Hotel. He was the chairman. In those days we didn't have charitable status, we were just a group who met once a week for a couple of pints in the Marine! We pestered the local MP and Tendring District Council. John was fantastic because on the surface he was a gentleman but he had another side. When you met John at work he was John Barker, owner of the business but when you met him when we were having a meeting he was a fire rag. He could certainly inspire people in his own strange way. I became involved because they knew that I was a writer and so they asked if I would write them a leaflet on the wildlife at the Naze. Somehow they managed to drag me on to the Save the Naze committee. Steve Humphreys, an amazing graphic artist, and I designed the logo and I wrote an article which the local paper produced as a two page pull out. We had to be careful though as there were a lot of characters on the committee, including a chap called Dave Hall who worked for a free local paper, based in Walton High Street (the Walton and District Advertiser). The pictures that Dave used were taken from the Observers Book! A lovely character though. Our secretary was Di Chate, who lived at Alfred Terrace. Di was lovely. She had worked in London for most of her life before moving down to Walton. She was the only person I knew who could drink me under the table pint for pint! Her idea of keeping minutes was jotting down this and that as she heard it and occasionally she'd fall asleep. So, in conclusion it was pretty informal!

BE: *Did that change as time went on?*

MCT: It did yes. We started to split in to two factions. There were those who wanted to work with the Council and there were the radicals. Sadly I didn't agree with those who wanted to work with the Council because every time the Council said we couldn't do something they said 'oh ok', then sat down and concluded, 'well, we'd better leave it then'. Nothing was ever done. I disappeared off the Save the Naze committee until it restarted as the Naze Protection Society in the eighties. I had a phone call saying 'Mike, we've restarted. Would you like to be on the committee or come on in an advisory capacity?' It went really well. We had our problems and our disagreements, although with any committee you get characters and strong personalities. I feel that what has been done at the Naze is good but to properly save it you have to drain it and sadly, nobody is up for it. We should not have let the Victorian system break down. If the stud breakwaters at the northern end of the Naze were still there then the problem would not be as bad. Longshore drift runs north to south along the east coast apart from a 1km section. From Crag Walk to the entrance to the Walton Channel sand travels northwards.

BE: *Do you feel that the cliffs' designation as a SSSI hold back protection of the Naze?*

MCT: Yes, but what we also have to remember is that the cliff is important from a geological point of view. It is the most complete complex of late Pleistocene crag anywhere in Britain. It's a hard one. Yes, the SSSI status does stop work from being done but we also need the statutory designation to protect the geology. It's a catch 22 situation unfortunately. I think that sometimes we have to look at areas and think that, sadly, we are going to lose it. Some people even say, 'well there's nothing at the Naze, why save it?' Well, if you open your eyes, there is so much up here. A few years ago we had a Bronze Age burial drop out of the cliff and there will be others. I suspect that if a ground radar survey was done quite a few burials would be found.

I can just about remember the pillboxes on the edge of the cliff. They were just on the edge when I saw them. I remember one of them with still a metre in front in 1962. What brings it home to you is when you see a large chunk go. About seven years ago I took a school group out on the beach. I came back the next day and we couldn't walk where I had been the day previous because there had been a big slump. I was actually on the beach fishing when that happened and the slump landed between two of us who were fishing there.

BE: *Today, over 100,000 people visit the Naze each year. How has that altered since the '60s?*

MCT: It's hard to say because in the sixties the cars were allowed all over the Naze so there could be as many cars as you would have on a busy Sunday today but they were dispersed everywhere. There wasn't a designated car park and the cars went where they wanted, including over the cliff! I suspect there are actually fewer visits today because holidaying at home is not what it used to be. In the sixties there were five caravan camps at the Naze alone. We lost two of those in 1963 when the flats were built at the bottom of the hill. There are local people who have never been to the Naze. I led a group here a couple of years ago and a lady said, 'I've lived here twenty five years and didn't know this place existed.' I couldn't believe it. She thought the Naze was the area leading up to the thatched cottages.

BE: *As a Waltonian do you see a difference between the Naze and the town?*

169

MCT: Yes although it's not so much now that we have lost the shops at the Naze. When I was a kid we had seven or eight shops. There was Blue House Dairies, Flettons, the butchers next to Flettons, Lilley's was opposite the Foundry Lane, the post office (now the Naze café) and a pet shop, opposite was Hipkin's take away and fish restaurant, and round the corner was the newsagents with a hairdresser above it. On the other side of the road was High Tree Stores. So, most people from the Naze (most locals) wouldn't go into town. They might go in once a week or once a month to get their hair cut but even for that we had a wonderful old boy called Fred Knights who lived three doors up from me who cut your hair for a shilling. Most people in First Avenue or Naze Park Road went to see Fred. He was actually a painter and decorator but he had cut hair in the Army for a penny a go and carried on when he came home. Also, on the corner of Florence Road there was a café which opened in the summer. It was a private house in the winter. Even amongst the kids there was an 'us' and 'them' thing. We'd meet on the border at White lodge Corner and if those townies gave us any trouble there would always be a fight! I remember a kid from the town coming up to see us and he was stopped by a few of the lads who said to him 'what are you doing up here?' as if he'd walked miles. As soon as he'd passed White lodge we thought 'you're on our territory now sunshine'.

BE: *Did many of the people you grew up with move away?*

MCT: A lot of them have moved away. Some of them stayed fairly local – Wix or Manningtree – but most of the people I went to school with no longer live in the area. I moved away for a while but came back and somehow ended up five doors down from where I was born. Walton is not the town I knew as a kid though. We've lost the shops, we've lost the banks and as one of my mates always says, we've lost our telephone exchange. For me it has changed beyond recognition. When I was a kid I could tell you the names of everyone who lived on Naze Park Road and the Avenues. I knew everybody. Now, I barely know my next door neighbours. It is sad. It has lost that community spirit, lost that sense of being. The heart has been ripped out, certainly at the town end but also here. The town has become so built up. I can remember when they opened up Gunfleet School as it was, now Tendring Technology College, walking across the fields home. Those fields are now houses.

BE: *How did you feel about that development?*

MCT: Well, it was rather sad because we used to play there. If we went over there at all then that was where we were heading – Turpin's Brook. Now there's a little bit at the back end of Mill Lane car park and another little bit between Turpins and Elm Tree Avenue but that's it.

BE: *Did you go out to the Backwaters as a boy?*

MCT: Yes. Every Sunday without fail we would go out to the old bird hide on the sea wall (which was burnt down about fifteen years ago) with John (Weston) and we'd swim over to Horsey as we knew the family who lived there. We'd then sit on the corner of Horsey and watch the world go by before swimming back. I look at that now and think how much closer it is today. I'll never forget one day I was half way across, grabbed hold of the anchor chain and suddenly things started jiggling about. All I heard from the guy in the boat was 'I can't get my anchor in'. I also used to spend hours wandering about on Hedge End.

I've led a charmed existence in many ways. The Naze has kept me sane over the years. It's strange but it's almost like a comfort blanket. It's very sad to see it disappearing so fast but it's got to go at some point. If we're losing land then somewhere else is building up. If you go down to the Kent

Coast you can see that the Isle of Thanet is actually gaining land.

2. John Fleming

One cannot ignore the rest of the local area when writing of the Naze. I was interested to hear about life on another local farm and so I spoke to John Fleming at Birch Hall, just outside Kirby-le-Soken, where he has farmed for most of his life, as his father did before him. I began by asking him to explain a little bit about his background in relation to the area.

JF: Perhaps we should start at the beginning. I was not born an Essex man but in Enfield Wash, Middlesex. The extraordinary thing is that my father was farming 300 acres in Enfield Wash – perhaps it is difficult to imagine a farm in the middle of Enfield if you know it now! Father had started farming in 1928 at College Farm which was owned by Jesus College, Cambridge and the College would sell off thirty or so acres most years for development. The farm therefore gradually became smaller and smaller. After eight years Father said to the College that the next time they sold off a bit of land they should take his notice as well. When they sold off a bit as usual he came down this way and looked at three different farms and decided that Birch Hall had the best farmland and so he bought it. Fortunately my Grandfather was able to guarantee Father's mortgage. We arrived here when I was two, so I am almost a life-long local.

When I was six my elder sister and I were compulsorily evacuated to North Wales and put on a bus outside Frinton gates in May 1940, complete with the rest of Hawthorns School. We saw very little of our parents during the next eighteen months. Petrol was rationed so they didn't have the fuel to drive to North Wales. I think they came over about four

172

times in the eighteen months we were there for about a week at a time. So that was my start to life.

My memories of Birch Hall (and the local area) are fairly sketchy because I didn't really grow up here. I do remember, however, that we had two army camps on the farm - one in the field just as you come in through the bottom of the drive and another one down beside Kirby Creek. The camps were because of the doodlebugs which came in over the farm on a certain flight line so it was sensible for the army to put up some gun emplacements on the farm. I remember the army camps because it was how I made my first honest penny. Field mushrooms were plentiful in those days and I would go out before breakfast with a bowl, fill it with mushrooms and then take it down to the cookhouse on the camp and flog them to the officers' mess. I remember on Easter Sunday of 1944, twenty three doodlebugs came over which the camps shot down all but one. Doodlebugs had a very distinctive noise. We could also see them coming as they had a flame coming out of them as they lumbered across the sky - and then all of a sudden BOOM – they had hit it.

BE: *How involved were your parents in the wider community?*

JF: My father was quite involved with the war agricultural committee. He was selected to keep an eye on local farmers because, of course, it was important that the nation produced as much food as was possible. He had to walk round a lot of local farms with the farmer to see if they were doing a good enough job. I was allowed to join the walk when I was at home from school on holiday. I can remember that quite vividly although where we actually went I cannot remember. My knowledge of the local geography was not good then, only coming home for the odd holiday now and then.

BE: *Birch Hall looks out on Skippers Island and I understand your father had stock there?*

JF: During the war every bit of farmland had to be farmed. I have no idea if Skippers ever had been farmed but my father was obliged to make use of it. Up until then it was just wild grass that was up to your knees. It was the horse and cart era still, not mechanical, so father decided that he would send about twenty heifers with the bull over there for the summer. You can imagine what fun it was trying to walk cattle across the hard! They would have a horse and tumbrel and would rope a couple of heifers on either corner and lead the horse across on the hard, which was fine – the mud on the hard was probably anything up to a foot deep. We had a staff of about ten in those days and every man was there to help. The rest of the heifers were pushed onto the hard but of course they did not stay on it. As you can imagine they were quickly up to their bellies. It was a day's hard work to get these twenty heifers and bull across onto the Island but it had to be done. There was then, of course, a repeat performance in September when they came home again.

BE: *What happened to Skippers Island after the war?*

JF: Nothing until the 1953 flood when the sea wall was breached and was never repaired. There used to be a hundred acres inside the wall at Skippers, a substantial piece of ground, but now there are only about twenty five acres – about fifteen acres on the east end and ten acres on what is known as the heronry on the west. Fred and Rowley Williams owned the island for many years before Essex Wildlife Trust.

BE: *Can you talk a bit about the beginning of your own farming career?*

JF: In my last term of school, when I was back for half term, my father said to me "I don't know what you are going to do when you leave school, boy, because I don't want you home on the farm you know". I wondered what I had done to upset him. He then went on to qualify that he had seen far too many fathers and sons go into business together and then they would fall out. The old man wanted to (*John pointed down*) and the young man wanted to (*John pointed up*) and then they would argue and squabble. So he said "You had better go off and make your own way in life and then when I've had enough of the farm I'll tell you and if you then want to come and be a farmer that's fine." So that was what I did and it was not a problem as I had a place at Cambridge University booked. Of course there was National Service in those days, which, as a farmer one could have avoided but I did not intend to. So, after school, when I was eighteen, I joined up in the Royal Engineers which was, without a shadow of a doubt, the most profitable years of my life. Having lived in a fairly sheltered, secluded environment at private school all my life, to join the army and go in as a recruit was a real eye opener. I learnt more about life in those two years in the army than any other period of my life.

BE: *You weren't tempted by a forces career?*

JF: No, although I really enjoyed it and I stayed on in the Royal Engineers, the corps that plays with explosives. The army would not train you as a national serviceman to play with explosives and as I rather liked that I actually asked to stay on for an extra six months; I never really did anything of much significance apart from involvement in the building of the M2. The headquarters of the Engineers is at Chatham and they were just making the A2 into a dual carriageway and as part of our training they sent us out for site clearance work. Mostly, it was trees that we were clearing before the soil moving equipment came in. They had this lovely stuff called 'Detonation cord' which was just like white electric light flex on a big roll. Seven turns round a railway line cut the line just as if you had cut it with oxyacetylene. You might have wanted two turns round a branch as thick as your thigh – your arm might have required one. There was a limit of not much more than about fifteen inches. After that it didn't actually cut, it just blew everything apart. We would wire up about five or six trees at a time. One turn round that branch, two turns round the next branch, three turns round another branch and when you had done that you just rolled it out to the next tree. We used to wire up six or seven trees at a time then connect it up to the button and get our heads down, press the button and then there it was – all laid as logs. Absolutely amazing and lovely fun! The trunks were then blown out individually with plastic explosive.

I then came out of the army in July to go to Cambridge in October. In September, however, my poor old father had a ruptured duodenal ulcer. They carried him off to the Middlesex Hospital in London where he stayed for six months. It was very touch and go. Mother said to me, "While the old man's away I think you ought to stay at home and look after the farm." That was the biggest giggle you can imagine, never having been much on the farm. But I couldn't very well say 'No I won't do it' so I came home for the year. Fortunately we had eight or ten staff on the farm in those days. We still had seven working horses looked after by Bert Madison, the head horseman, who worked with father for thirty odd years. Bert came at six in the morning and the day staff at seven so I used to go and see Bert first thing. "What are we going to do today Bert?" I'd say. Bert would say "Well I reckon we ought to do so and so" and this is, of course, what we did.

Eventually father came home again but he wasn't really in a fit state to do too much, so I saw the year out. Father obviously had a change of heart – he thought that having me at home perhaps wouldn't be the disaster that he had thought. So he asked if I would like to join him on the farm. I said "Yes I'd love to" and he said "Well you don't need to be a graduate to be a farmer!" so I had to tell Cambridge that I would not be coming to see them after all. Father then said "I think you ought to go and learn a bit about farming" so off I went to Writtle for a year: the biggest waste of time you could almost imagine. Everything I learnt there I could have read in a book in a fortnight and I spent a year doing it. Lovely fun though! I hardly had to work at all and I played a lot of sport. After that I have been farming ever since.

BE: What were the enterprises on the farm at that time?

JF: When I was a boy we had cows, pigs and sheep as farms did in those days but there were no pigs and sheep by the time I was married in 1960. When I came back from Writtle, Birch Hall was about 400 acres, not a very big base for two families and Kings Farm came on the market. I thought it was just right and we needed to buy it. So I said to father that we needed to extend our business to bring in two reasonable incomes. Father would not hear of it. "No, no, I'm not going to borrow money" he said, "If you borrow money then you go bust". He had started in business in 1928 in the 'depression' and if you borrowed in the 'depression' then you went bust. I could not change his mind. My mother was a bank manager's daughter, not brought up in agriculture at all and she understood finance much better than my father, but even with a bit of help from her we could not persuade him. He was not going to borrow money. He had obtained a mortgage on Birch Hall and that was enough. I said "Well, we have to jig the business up somehow" so we started expanding the dairy herd from 40 cows to 250 cows plus followers and calves and eventually we became a big dairy herd, one of the ten largest in the country. The irony was that we borrowed more money to do that than it would have cost us to buy Kings Farm. The cows did us very well. We had three full-time cowmen, plus myself, and needed two on at any one time, so we were still able to run a five day week as did the rest of the farm staff.

BE: 'What are your memories of rabbiting/shooting?'

JF: There were young Fred and old Fred Pratt. Old Fred was quite old in relative terms because young Fred was about thirty. I remember young Fred particularly well. It was the days before myxy (myxomatosis) and they used to shoot rabbits all through the winter at least three days a week. You had to keep on top of them. It was my job as a boy to scoop up all the dead rabbits as they were shot and pull them home in a wooden bushel box which I had mounted onto a pair of wheels. On Monday morning all these rabbits would go to Thorpe station to be put on the train to London. All the local farmers did the same. There would be up to six or seven hundred rabbits on this train every week. Marsh House and Birch Hall were run as one shoot, originally by Woodruffe (*J. W. Eagle*). Father was quite keen on his shooting but the Pratts (through Eagles) were always the gamekeepers. I am sure father did not contribute at all. Woodruffe would have done that. I think he was pleased to have Birch Hall added to his shooting empire (it was an empire to him and he was the emperor!).

BE: Do you have memories of Woodruffe?

JF: Yes. I remember Woodruffe and Frank *(his brother)* very well – they were very different characters. We acquired Batts Hall Field through Woodruffe. Batts Hall field is our furthest field going towards Landermere. It had not been cultivated for many years and the scrub was as high as this room *(10-12 feet)*. In the wartime, when every bit of land had to be farmed, the government got on to Wooduffe and said, "Come on, you have to farm this land". In the days of horses and carts it was a fair little trip from the Ashes *(on the Kirby/Walton boundary)* to Landermere to farm a twenty six acre field. Woodruffe came to father and asked if father would farm it? Father said "there is only one way I'll farm it is if I buy it". He said that if he had to clear it he wasn't going to do all that without owning it. Woodruffe probably did not have any option. He had reputedly won that field in a game of cards! I don't know who owned the field before Woodruffe but he was given that particular field because it was the furthest field away from Walton. He sold it to father at £8 an acre!

It must have been one of those lovely dry March afternoons with a good east wind when father burnt it. It cleared all the brambles and we were left with just the trunks. Then Rowley Fenn who had thrashing drums and steam engines, brought in a socking big cultivator. It just went backwards and forwards between the steam engines, probably about that deep (John indicated about two feet), bringing up all the hawthorn roots. Then all that summer, whenever they had a bit of time, they would pull them out and make big bonfires and that is how it came back into cultivation.

BE: Do you feel that Frinton and Walton have changed during your lifetime?

JF: Frinton hasn't changed hugely in my recollection. Connaught Avenue was always there – the Bond Street of East Anglia, which perhaps it is not now. The other avenues were there and on the left hand side, Pole Barn Lane. Inside the gates Frinton has changed very little except for in-building. Kirby Cross used to be just the Cross at Halstead Road. That is what has changed of course. As you went through Kirby Cross there was nothing on the right towards Frinton, or at least very little. The lanes on the left were there. It is that development between Turpins Lane and Elm Tree Avenue that has made the most change. Meres Farm was still there when I was married. The land is now built up all the way from Kirby Cross to Walton. Lower Kirby has changed with houses along from the church now where there were just little shacks and the development along Halstead Road was just fields. The forge in Kirby lasted for quite a long while.

BE: What about sailing days, boats and the yacht club?

JF: I grew up into boats because I suppose my only boyhood friend round here then was Dennis Cross. Colin Cross (Dennis' father) was commodore at the yacht club. Dennis was encouraged to sail and was a very good sailor – he was very good at everything he did. I used to crew for Dennis and that is how I started into sailing. I really enjoyed my sailing with Dennis, it has been my hobby all through my life. My first boat was a Jewel – a Walton One design. When we had children we bought a 22 foot kestrel but it soon was not big enough. Then I spoke to Ron Wyatt from 'Bedwells' in Mill Lane. I had great admiration for Ron – we were in the Rotary Club together. Ron was about ten years older than I and he had been in the war and was always good company. In 1973 I said to Ron "I have to have a bigger boat Ron, we have outgrown the Kestrel now'. He said, "There's the boat for you" (pointing across at Widgeon in a mud berth just across the creek). I thought that if that is what Ron says, he knows far more about boats than I do – so I bought her and she is still down there in Kirby Creek to this day even though she is 113 years old and still in very good condition. I have no intention of parting with her. I have had my money's worth out of her many times over. I have sailed her around the UK twice, over to Ireland and up through the Gota Canal in Sweden, into the Baltic, round Brittany and through the Dutch canals many times. I joined the yacht club when I was fifteen and I am eighty now so I have been a member for sixty five years. I can remember when David Blyth was thinking of building a marina. He approached Ron Wyatt and John Halls before he went to John Titchmarsh. At that time John Titchmarsh was only building dinghies and little boats in Churchfield Drive in Walton so I can understand why he jumped at the opportunity. Hamford Water (or the Backwaters) is a very special area and quite rightly designated a Site of Special Scientific Interest. Bridgette and I are very honoured to have our kitchen window looking north down on it from the great height of 53 feet.

3. John Halls MBE

John Halls is well known in Walton and was made an MBE in 2009 for services to the town. He has previously served as a town councillor and as the Town Mayor. John was born in Walton in 1933 and has lived in the area all of his life. The Naze is directly linked to the town, administratively as well as geographically and I was interested to find out more about twentieth century Walton life and particularly how the town has changed. I began by asking him about his early life in Walton.

JH: I was born in Walton on 12th April 1933 and have lived there all of my life. My mother always told me that there was so much snow on the ground in April of '33 that Dr Brockwell and Nurse Watts couldn't get to the house to deliver me. Perhaps the first thing I can remember as a local lad was a party at the old school in Walton High Street, which is now the library, when I was about five. My brother Derek took me along. In those days it wasn't called a party. It was called a 'bun fight'. I went home afterwards and I was crying and my mother said 'whatever's wrong?' I said 'well mummy they didn't throw any cakes or buns at anybody'. That is my first memory in Walton. I started at Walton Primary school in

Easter of 1937. I can always remember my first day at school as I was crying my eyes out but I soon got used to it. I can also remember one instance vividly when I was walking down North Street hand in hand with my brother, past Wrights the bakers (they must be fifth generation now). All the parents gave their children a ha'penny and you got two bread rolls for a farthing each so both my brother and I had one. They were lovely rolls. Another memory I have of school was when I was in a class with Miss Lyers. There was a lad there, Alistair, whose father was the local policeman and I turned round to speak to him in the class. Miss Lyers grabbed hold of me and in those days you wore little shorts. She smacked my bottom hard. When I went home I told my mother what had happened and she took my trousers down and smacked me a good sight harder! That comes over very vividly. When the war started a terrific number of children in Walton were evacuated, mostly went to Stroud in Gloucestershire, but we stayed on and were schooled by various teachers. Miss Grey I remember very well and a Miss Modlen. We went to various peoples' houses for various lessons – arithmetic in one house and then for English you went to Miss Modlen. There's many a child who could remember being smacked by Miss Modlen. The boys that stayed during the war included Ronnie, Don Hipkin, my brother Derek, Alec, Colin and Ronald Shaw, Smudge Smith as we call him, John and Bryan Hicks (although they were evacuated for a time) and many others. One Sunday night in 1942 we were in Walton Church and the German bombers came screaming over, machine gunning everything. We ran out of the vestry, Ronald Shaw and myself, we ran up Martello Road as a bomb came screaming down and it dropped right on top of the police station and the house next door where Mr and Mrs Catter were. They were both killed and the police station flattened. There was also a Catholic Church next door which was completely flattened. In those days there were lots of trucks up at the railway station which was quite busy at the time because all of the farmer's milk went away by rail and on that particular Sunday night one of the bombs bounced and it hit the railway track, then went over the top of Twizzle in Mill Lane where Mr and Mrs Briscoe lived. It then went right into Major Cripps's house, killing Major and Mrs Cripps. Ronald and I had run down Martello Road, threw ourselves down alongside the hedge, then ran a bit further and these planes were still coming over. Of course when we got to our houses the fronts and backs were all thrown out. Jumping ahead, when the doodlebugs started coming over there were gun crews here. During the day they brought the guns up and hid them on the Martello field. At night they would take them over on the marshes and wait for the doodlebugs to come over but this was a Sunday evening and at about 6.30 or 7 o'clock the doodlebugs were screaming over. These chaps, bearing in mind there weren't any houses on Mr Blyth's farm or on Mr Eagle's farm back then, brought this 3.7 gun down so low, firing across the town, that they blew the public bar at the Albion Hotel clean off. Fortunately the bar wasn't open at the time. There's another story from early in the war, I think in 1939, when a Japanese boat was off the coast, carrying general cargo. She hit a German parachute line and blew up and a lot of stuff washed ashore on the beach. Ted Oxley and my father retrieved an awful lot of stuff because only the lifeboat crew were allowed on the beach. At that time live wire was going up. They'd got a load of broom handles and one laughs when you see 'Dad's Army' today but it was so true. These broom handles were in bunches of thirty six and Mr Oxley and my father sold these broom handles to Essex County Council for a farthing each. They were then given to the local Home Guards. There were also sheets and sheets of very thin plywood and they sold this to Greens the builders and various people in Frinton when the soldiers took over all these beautiful houses in the Avenues. They used to cover all the floors up so the army didn't damage them. The local lifeboat crew: Mr Bloom,

Jonas Oxley, Ted Oxley, my father, old Bunker Cooks, Mr Frasier Baker and Mr Claude Brooke etc relied on us to run round knocking on the houses saying 'LIFEBOAT, LIFEBOAT'! The other member of the lifeboat crew was a man by the name of Toddy Williams and he was a yeoman of signals in the Navy for many years. He was the Signalman on the Lifeboat. Of course it was all done by semaphore or with the aldis lamp flashing. The lifeboat was then in the river because they'd blown two great big holes in Walton Pier so that if the Germans landed they couldn't get up the pier. The sea wall was flatter in those days. As a boy I would ride all the way round to Colonel's Hard and say to Mr Brooke or Mr Oxley that I had a message for them that they had to go somewhere or do this etc. The other wartime story is about the boating lake. They kept it full during the war and there were two areas, one of which was deeper, in case there were fires. They could put suction hoses in there to pump the water out. On one particular Saturday at about 4.30pm we had a tremendous thunderstorm and the end of Walton Pier, which was then a lovely pavilion and dance hall, was struck by lightning. They pumped the water up from Mill Lane to another pump and then up Newgate Street on to Walton Pier, right to the end and managed to fight the fire.

Anyway, time went on and when I eventually came home from college my father said 'I've made you a bound apprentice at James and Stone Shipyard at Brightlingsea to learn the business.' I said 'Dad, I've had no holiday'. He said 'You're not getting holidays. I've bought you a saw, a plane and a two foot rule and you're on your own.' I was then sixteen and that was a bit of a shock. I stayed at James and Stone for the next five years. I caught the 6:04 train to Brightlingsea every morning and in those days you went to Wivenhoe, changed at Wivenhoe, caught the Crab and Winkle to Brightlingsea and then you had about a ¾ mile walk to the Shipyard. If you got there at 7:30 and clocked in you were ok. 7:31 and you lost fifteen minutes of wages. My first pay packet was twelve shillings and sixpence. After a few months they put it up to fifteen shillings. I worked in the shipyard until I was twenty two because on 13th October 1953 I was called up in the RAF for national service. I went to RAF Cardington to sign on and after that I was posted to RAF Hensford in Staffordshire. I was sent there as potential officer material and was there with about 6000 people on the camp. In December, at five or six hundred feet above sea level it was damn cold. Unless you got to the wash house before five in the morning you didn't get warm water at all. We finally passed out of there in December 1953. I came home and I weighed 11.5 stone then, never fitter in my life. They should never have gotten rid of national service. I'm proud that I went into the RAF and I had a wonderful time but I suppose you always think of the good times and not the bad. I came home on leave and my posting came through to RAF Mountbatten in Plymouth. I was known as a Boatwright and I was posted in the Marine Craft Unit. They told me that I had to take a six week exam. I met up with a chap called Laurie Holder who came from St Marys on the Scilly Isles. He'd never been to the mainland and was a lovely guy. We went up to this chap Warrant Officer Cocker and we said that we'd both just done a five and a half year boat building apprenticeship so we're not going to sit down and do another exam. He said that we were a couple of rebels. We said that we weren't but we just wanted to get doing something. I remember that instead we did a direct entry test in which we had to take a plank out of a Whaler and put a new one on the turn of the bilge. There was also a written paper which we each did in about ten minutes and that was it. We were both sent up to RAF Calshott and I lost sight of him then. I was then put on a thing called a SRIM (Special Radio Installation Modification) and I travelled to about fifteen different camps all over England. I went to RAF Bridlington, RAF Driffield,

185

RAF Coningsby, to Boston in Lincolnshire, Porthcawl in South Wales. When we were there we had to go across the Bristol Channel to target tow for the RAF regiment. The skipper said to me about 8:15 in the morning 'would you like a brandy?' I said 'No, not at this time in the morning.' But, the target was about half a mile behind and as the shells started dropping round the boat I asked him if I could have that brandy! After several years I was posted back to RAF Calshott full of complaints. They said 'Halls, you've got so much to say that you can represent C squadron when the AOC (Air Officer Command) come for inspection. So, cutting a long story short, when the AOC came the old boy sat down and said 'sit down Halls. I'll look at your record etc' and he said 'I see you come from Walton-on-the-Naze in Essex'. I said 'yes' and he leant forward and said, 'tell me, do they still sell ice cream on Walton pier?' I said 'yes sir' and he said that in the late twenties his parents had used to take a house in Frinton for the Summer and his nurse used to wheel him from Frinton to Walton to get him an ice cream. Of course, these poor devils were still standing to attention and we were talking as if we'd known each-other for years.

BE: *Returning perhaps to wartime Walton, how much of a presence were the army for the local community?*

JH: Well, of course there were soldiers posted at the Naze and so local people weren't allowed on to the Naze, only the Eagle family and the farm workers who would go up to the Hall where they had cattle. The Naze Tower was a radar station and on the right hand side of Hall Lane there was a field (today a children's play area) where all the search lights were. The radar at the Tower was all tied up with Bawdsey radar. In Frinton, in front of the Grand there was a great big gun. When they started firing it the whole place shook. Also over the bridge by the Masonic Hall there was a great big Army camp

186

with lots of Nissen huts. There were an awful lot of troops about here what with Harwich being a very important area. The lifeboat was out seemingly all the time, whether that was picking up German pilots or English pilots. They were kept very busy.

People talk about the Naze cliffs falling at the moment but they didn't fall much during the war because right the way round the country there were defences, scaffold poles all criss-crossed together with barbed wire in between them. These broke the force of the tide so the cliffs didn't continue to fall until after the war. The two pill boxes which are out on the Naze beach today were still around a hundred yards in from the edge of the cliff. I have photographs to prove it. After the war the Forces captured a lot of films from the Germans. Believe it or not they got aerial photographs of the Naze, of Harwich, of everywhere. We didn't know what was going on at the Naze but the Germans knew.

There was also a beaufort gun on top of the Martello Tower because the Martello camp was used as the cookhouse for the Forces. They used to march down Mill Lane right up to the Martello camp. There was another Sunday evening when they lit the entire town up with incendiaries. They dropped them all over the Martello. We've had our share of war at Walton-on-the-Naze I think. When the war ended there was a great big prisoner of war camp over at Ramsey. All the Italians were sent all over the area to take the structures down and clear it all up. There were however still lots of mines. After the war lots of dogs hit mines, when chasing rabbits and that certainly did some damage to the cliffs. We didn't know anything about D Day and one morning every road in Frinton and Walton had lorries and troops and guns but it was kept so secret. They were there on 5th June but on 6th June there was nothing left.

They'd all gone overnight, ready for the invasion. After that, the whole lot quietened down in Walton.

BE: What did Walton High Street look like at the time?

JH: Looking from the Albion down towards the Church there was the Albion Hotel on the left and the Queen's Pub, then Mr Cresswell lived in the big white house. Then there was Dr Johnson. We had four doctors in Walton then: Dr Johnson, Dr Brookwell, Dr Dee and one other, all who had their own dispensary so they made their own pills and potions. Then there was Mills grocers and the Vicarage. The vicar here at that time was the Reverend Owen. His wife was a German and everyone thought, including the police, that she was going up on the allotments at night and flashing her torch. Then there was the post office, built in 1936 with a flat roof, with intentions of putting the telephone exchange on top. Next door was Mr and Mrs Drinkwater. The telephone exchange was under their house. After the war they decided that the telephone exchange should be in Frinton but there were very few phones. I remember that my father was number 96, Harry Barker at the Marine was 100 (he wanted that). Miss Moss, a lovely lady, was the person on the telephone exchange. She was about 97 when she died, never married, looked after her mother and she would know everything that everyone was doing. She would say 'oh, they're babysitting tonight or gone to the cinema so she wouldn't put you through. It was lovely in those days. Everybody knew everybody else. By the vicarage there was a piece of land called the Glebe land and there were tennis courts there before the war. With 'dig for victory' the lifeboat crew grew a lot of potatoes and distributed them round the town. We really had a good time then. Nobody had any money but everybody made their own fun. I remember when you had a radio in your house everybody would walk through the town with an accumulator to take to Mr Harvey at the garage in the High Street to get them charged up. You would screw it down, turn something light on and sit there listening to the radio on a Saturday evening. There was a programme called 'In Town Tonight' which everybody listened to and they used to say 'good evening, this is In Town Tonight. People have come to England to be in Town Tonight and they have travelled to England by land and sea', not a question about air, just land or sea. We also used to have quite a lot of important people come to stay in or near Walton, in Frinton at the Grand Hotel or at the Queen's Hotel. During the summer season businesspeople would come down from London to take a house at Frinton or stay at the Grand in Frinton or the Marine in Walton and their wives and children would stay for another five or six weeks. The gentleman of the house would commute to and fro to London and back. They were known as the 'August visitors'. They would come down and have a cricket match against the lifeboat crew or the Frinton tennis club or the golf club. If the tides were right they would go off and play cricket on the Gunfleet sands. People like the Gummer family, the Robinsons and the Bowsteads decided after the war to come and live in Walton and Frinton.

BE: *Can you tell me about your memories of the lifeboats?*

JH: One of my earliest memories is from 1940 when we had a boat here called the EMED. But I remember my father talking about previous lifeboats, especially the James Stevens. My grandfather, Frank Halls was a crew member in 1889 in the old rowing sailing boat which they rebuilt. They eventually decided to put an engine in her and my grandfather was the engineer. That went on until the beginning of the 1930s and then they built the modern lifeboat EMED and she carried on through the war. She was a lovely old boat but there was no cabin or anything like that. There was one bad night in 1940 with lots of ice and snow and when the crew came ashore Mr

189

Brookes actually died soon afterwards as he was so frostbitten. When my father came in my mother (we had a galvanised bath with the old days of heating the copper and putting the water in) made sure he had a bath. He was sitting on the chair with his water boots on and his feet in the water to thaw him out just so he could take his boots off. That was a very sad night. Time went on however and the new lifeboat to come was the Edian Courtauld. One Saturday night there was a big yawl and the chap came on deck and said 'coxswain, could you come aboard, the governor wants to speak to you.' Jonas Oxley, the coxswain, went aboard and in to the cabin and saw none other than August Courtauld. He said 'You'll keep this quiet but I am August Courtauld, the Chairman of the committee of the management of the RNLI'. The Courtauld family put up the money for the new lifeboat. They gave me leave from the air force to come back to Walton to see it christened by the Duchess of Kent at the Albion Breakwater. We then got the City of Birmingham and that was used quite a bit because in those days people didn't have the navigation equipment and people were always getting in trouble: most of the time it was rubber dinghies. In the Winter of 1962/63 the sea froze about a quarter of a mile out. The lifeboat was solid in ice. We weren't able to get to the gentleman on the sunk lightship. There were eight in the crew then and you used to do a month on and a month off. This chap was the lamplighter at the top of the mast and during that winter he fell off on to the deck and broke his back. There was nothing at Harwich or Felixstowe in those days and we obviously couldn't launch the Walton lifeboat. Dr Johnson went to the coastguards and told them over the radio what to do for the poor man. I think it was about five days after the accident that they finally managed to get a boat out from Harwich to rescue him. That was a very nasty one. There was another occasion when the boys went off to the Rough Towers and the Sunk Towers and the men on there were shooting at the lifeboat crew with rifles. The Navy was called in to sort it all out. It was just after that that they decided to do away with the Sunk Tower and the Rough Towers became the Principality of Sealand. The 50s and 60s was probably the busiest time for the lifeboat as lots of people had bought new boats...things have certainly improved when it comes to lifeboats.

BE: *What do you remember of the '53 flood?*

JH: I remember it well. There was a chap called Dave Dunn and we both had motorcycles. We went down to the boatyard to put them in one of the sheds about midday on the day before the flood. It was blowing quite hard and my old man said 'do you want to come down to the yacht club and have a drink?' So, we went up there and we looked out of the window. He said 'there's something brewing. The tide hasn't come up today'. Nobody believed him but they asked if they could put their cars in his yard. He told them to take them up to the railway station but they laughed at him. My father went down to Brightlingsea and at about 6.30pm the tide was lapping over the quay. We went in the evening to Molly Harmer's 21st birthday party at the Women's Institute. In those days you had to stop playing music and be out of the hall by 11:45pm. We went back to the Harmer's house and Mr Harmer said 'oh John your father's trying to get hold of you. He said that there's a big tide and he thinks you should go home.' Norman Weeks, who was then engaged to Molly Harmer and was one of the managers at Harmer's Foundry, had his mother's car – a Ford Prefect – PNO444 (I shall never forget that numberplate). The four of us – Betty, myself, Norman and Molly – jumped in and headed down to Walton Church. There was a bloke there with a torch stopping everybody. I said 'what's going on?' He said 'well that's the bloody tide there' (the tide was right by the lampstand in Walton Church. We saw my brother Derek and old Tim Oxley and Jonas come rowing up the road. They went up Kirby Road, taking people out of their bedroom windows. You must remember though that although it was high water at Hunstanton and Cromer and so on there was no television and nobody phoned each other. Nobody actually drowned in Walton but several had to be rescued – particularly in Churchfield Road. Some of the older people died soon afterwards of shock. It did a huge amount of damage. All the sea walls gave way and in Walton High Street the next morning there were horses and cattle and all the coal from Hipkin's coal yard in Mill Lane had gone everywhere. What people don't realise is that the tide came up the Pier slope and over the top of the sea front (as well as at the Naze and the Backwaters). The tide didn't go out either, because it just couldn't get away. There was then another very big tide in the afternoon. The Lady Eileen, the houseboat, had broken adrift and gone right across the sea wall. I remember that my father-in-law said 'why don't you get a fire pump to wash the water away' and that's exactly what they did. Then they lifted her and moved her back. Of course they had to build the wall up again and decided to build the new sea walls. I spoke to Mr Rollings a few years ago, he was then the engineer for the Catchment Board and I asked him where he got all the clay from? He said 'we bought it from Mr Eagle at the Naze'. Down on the marshes the lorries went backwards and forwards for months and cranes would lift each bucket individually to be placed by the engineers. It was really fascinating. So, a very scary time but we got through it.

BE: *Can you tell the story about the Town Hall in the High Street?*

JH: It was a great shame because the Town Hall was pulled down about thirty years ago. When the Council amalgamated and used the Council chamber at Frinton the Town Hall in Walton became redundant. It was at first turned into a restaurant by the Hipkin brothers which was quite successful. Then a chap named Geoffrey Bowett bought it and took all the slates off the roof and blocked the windows up. Of course the children then got around there, threw bricks and broke the windows so all the damn pigeons got in. Then came the rats and it smelt like the devil. Bowett eventually sold it for bricks and they pulled the whole place down. It should never have happened though. It was a beautiful building. I remember it very well as a youngster because they used to have a lot of functions there. In this day and age it would have been wonderful for the town.

BE: *When did you first become involved in Walton politics?*

JH: Well I met my wife Betty when we were sixteen, playing tennis, and we then both joined the Young Conservative Association and have been Conservatives for all our lives. I've been a councillor for about fifteen years. What I enjoy is working for the town, helping people and taking up their causes. I believe that when you are elected you are there for the people of the town and you should only have their interests at heart. I'm a strong believer in the town and I've tried to give to the town all my life. I've thoroughly enjoyed it whether it's been sailing, at the yacht club, the carnival, the lifeboat crew (I was for forty years in the lifeboat crew). I love meeting people, talking about Walton and helping all I can. I think that some politicians are far too serious today. I often think what Macmillan and Harold Wilson would think today at the way that some politicians carry on. But, I suppose life has changed and you have to change with it.

BE: *Do you think that sense of community used to be stronger in Walton and do you think there has been a divide between people at the Naze and people in the town?*

JH: In the 1950s and 1960s there were the Naze people and the Town people but that era has now gone and we work closer together. I think that's probably because all the shops have gone. In the sixties there was Emersley's the chemist, Mark James the newsagents, Mr Gooks the baker, Mr South Blue House Dairy, Mr Calver the butcher, Mr Agnes the greengrocers etc. Now there's just really the Naze Post Office. Everybody now has to come down to the town for their shopping. The caravan camps have certainly made a difference. Mr Garney owned Willow Caravan Camp and Mr Snare had one which Taylor Caravan Parks bought – then quite a few of the parks amalgamated. Mr Garton started the Martello and then Brian Watling took it over. Then there's the Naze Marine which has changed its name several times.

In those days everybody who had lorries – the Hipkin brothers, Grants, Moyes etc - would loan them to people for the Carnival. For the Lifeboat we would build our own float down at the boat yard. There was also a traditional flour fight. You would go up on the allotments for soot. You would then go to Mr Wright or Gooch and get a sack of flour and then go to Mr Wilson and get a load of sweet bags and bag them all up. Then the fight would take place on the Pier beach. It's a job to get that sort of thing to happen today. I think though that with the Walton Trail walks now and other things happening at the Naze we will be able to bring more people in to the town which will help it to be used all the year round, servicing the cafes and restaurants.

BE: When did you first get involved at the yacht club?

JH: I first became involved in 1946 when I was just a young lad. I used to go crewing in the days before the Jewel class. There were six of the eighteen footers – Jelasma was number 1 which belonged to Captain Cook in Frinton, number 2 was White Heather owned by the bank manager Mr Seabrooke, number 3 was Cinders, number 4 was The Surf, number 5 was Pierce-Eye which was Eagles, Bill and Bill were number 6. My father used to tow the eighteen footers to Harwich to race and they would race on Saturday and Monday because everybody went to Church on Sunday. In 1948 I went to James and Stone where I helped to build a lot of the Jewels which are still racing today. When they were built, dear old Robbie Stone, who designed them, would ride round on his bike and shout 'get that put right!' if there was an issue. They were £125 each and he reckoned that if we broke a plank or anything that was his profit gone. They went all over the country. I wasn't allowed to join the yacht club though because I worked as a professional boat builder. I got so annoyed because Teddy Carter, who had also been a boat builder, was allowed to join the Club. It was through Clifford Green, who was then Commodore that I was allowed to become a member. He said 'this is ridiculous. Young John builds the boats and the Halls family do a lot for the Club' so they scrapped the by law and I could join. However, I couldn't helm. After the eighteen footers, Brooke and Halls built the gem class boats which were lovely sailing dinghies. When Mr Brooke decided to pull out to start an ironmongery business my grandfather bought him out and it became Frank Halls and Son (my father), which it still is today.

BE: *Can you tell me about your memories of the Mere?*

JH: The Mere is very sad. In the 1920s Brooke and Halls bought all the land down there. The road to the quay was £12, the saltings were about £14 and the quay itself was £11. In 1921 they were offered the Mill Pond and the asking price was £60. They couldn't however stretch to it. A chap named Josh Francis – Francis and Gilders – bought it. Josh Francis's daughter married Mr Carter and they turned it into the boating lake with lots of little sailing boats. When Mark Carter was alive, if someone turned up at 9pm and wanted to go rowing he'd get a pair of oars and give them a boat but he died in 1948, aged only about 54. He had a heart attack. Ted Carter then ran it and it went well but all of a sudden he decided that he was going religious and he wouldn't open on Good Friday, then he wouldn't open on Sundays and eventually any religious festival. He went in to his shell. We tried hard to buy it two or three times and we tried compulsory purchase. We formed a company with Nobby Pearce and his son John, my brother Derek, myself and Ron Wyatt. Nobby Pearce and myself went to see David Brookes in Colchester – it was January 1972 – to sign the contract and we needed to pay capital gains tax. Ted Carter was sitting there in a white mac, we sat down and he said 'gentlemen, I've given a lot of thought to it but as long as my parents' names are so dear to me I won't sell the lake', then he got up and walked out. We couldn't believe it. We'd agreed everything and we were going to pay. So, we formed another company later – myself, Brian Watling, John Neegis, Chris Porter and a nephew of mine – got it all set up again and it was going to be compulsorily purchased on the Monday morning. We used solicitors in Manningtree – Cox and son – and drew the money out of the bank (it had to be cash). The Council (it was a Labour Council then) held a meeting out at Ardleigh on the Friday and unfortunately it was lost by one vote. So, twice we tried compulsory purchase but you couldn't talk to Ted about it even though we used to go drinking and he was one of the boys. It was a great pity though – I often wonder what Mark Carter would think if he could see it now.

BE: *You are a trustee of Herbert Columbine's Victoria Cross – can you tell me how you came to be involved so centrally?*

JH: Mrs Columbine left all his medals to the people of Walton. They were put on show at the Royal British Legion from the war to 1964 when the local greengrocer, Charlie Dean, who was Chairman of the House Committee at the Legion, said that we couldn't keep the medals at the Legion as they would be stolen. Michael Hipkin who was one of the trustees along with Charlie Dean and Mr Snare decided to get replicas made and put the VC and his other medals into Westminster Bank, where the Royal British Legion banked. I was made trustee in 1964 when Mr Snare died. They were then moved to Barclays Bank in Walton and then Michael Hipkin died and we were looking for another trustee. I suggested David Eagle, as he's a local man with sons etc. I believed the trustees should remain locals. When Charlie Dean died we asked Chris Brookes to become the third trustee. Mr David Eagle did a lot of work and we had people writing to us saying they were Columbine's second cousin and so on but after discussion we decided that they should be put on show so that everyone could see them. We contacted the military museum at Chelmsford and David and myself went along to the bank to collect it, before handing it over to the police. It then went in a police car to the British Legion at about 2pm. We then handed it over to the curator of the museum at Chelmsford (by that time we knew it had been insured etc) and we've signed it away for '36 months'.

4. Bryan Hicks

Bryan Hicks has also lived in Walton for most of his life. Born on the Naze in the 1930s Bryan has experienced much change in Walton over the years, has memories of wartime Walton, despite having been evacuated to Gloucestershire early in the war and has immersed himself in the local community, serving as a councillor and as secretary of the YMCA for many years. I began by asking him about his early memories of Walton.

BH: The first thing I remember is living at the Naze in a big house called Ayrthwaite, which was a boarding house. We used to have up to forty people there. Directly after the war, when there was rationing, we fed all forty people but they had to stay in various houses around the Naze because we couldn't sleep forty people. We had nine bedrooms. Both before and after the war on Saturdays my brother and I had to take the cases to the various houses. Unfortunately, in about 1955 our house fell down. It had been built on top of a hill without proper foundations. We had some light guns in front of the house during the war together with search lights and they eroded the land in front of it so it had to be pulled down.

To get to school we would sometimes walk along the beach, sometimes down Hall lane and sometimes down by the marshes. We would often get in to trouble for having wet feet. My brother and I usually had to help some of the other children. There were a lot of very poor children in Walton and for some reason we used to help with the maths and English and yet we were only eight or nine. The school in Standley Road (the junior school) had a wonderful playing field and I think everyone remembers the smell of the grass when it was being cut. We were only in the junior school because we were evacuated when we were nine and then when we came back we went to Colchester.

BE: *Who were your teachers?*

BH: There was Miss Jones the headteacher at the infants school and Miss Lockyer who came from Thorpe (she was a wonderful teacher). During the war they used to bring us 1/3 of a pint of milk which was always warm – horrible. We also had malt in cardboard containers – it was supposed to be good for you. There must have been at least 150 children at the school, perhaps 200.

I remember directly after the war the Borstal Boys used to come down from Walton and we played football against them on the Martello Camp. When I was about 20 or 21 we would always play the 'visitors' on the beach adjacent the pier when the tide was right for charity. There were always hundreds of people there watching. There was an article written by John Flint, who used to run High Tree Stores at the Naze and later moved to Australia. He reckoned that my brother and I were snobs and yet we weren't like that at all. We used to enjoy playing on the marshes and swam at Foundry dock and out to the Twizzle on a high tide. It would be up to twelve feet deep.

BE: *How do you remember the High Street?*

BH: The High Street was absolutely wonderful. It had everything that you could possibly want – for example we would go for Munnings for fish - and at Christmas time it was beautiful, particularly before the war. In the summer it was packed. I remember that on Saturdays my brother and I would put leaflets out for a man called Clown Sunshine (Joseph A Cookson) who used to do shows at the round gardens and on the pier. He would give us sixpence to post leaflets to various houses and the tents opposite us – the area opposite Third Avenue which is still an open space and then called Jubilee Camp. There wasn't really any other way of earning any pocket money. When the war started I remember vividly the pier being blown up in two places in March 1940. They thought that the Germans wouldn't be able to land if you did that. I remember standing at Jubilee Camp and watching it happen. During the war we had thirteen airmen stationed in our house. My dad had to go to London as a fireman and my mum stayed behind and was paid £1 each by the airmen to feed them and do their linen etc. There wasn't much schooling in Walton during the war and my brother and I went to Colchester. The train fare to Colchester (I went to the Boys High) for a quarter (3 months) was about £3. There was the old steam train and the turntable where the car park is today.

BE: *When did you first hear that you were going to be evacuated?*

BH: It must have been around March 1940. We were fitted with our gas mask and packed our suitcase with our belongings and off we went – it took all day. When we arrived at Minchenhampton we were put in a room above the market square and were selected. John and I were the last to be picked – I suppose because were red headed twins they didn't want us. The two Borons – Fred and Arthur – their father was killed in a Japanese prisoner of war camp – came with us and we stayed there for 2 ½ years. They were a reasonably well to do family. There was an old lady, her daughter and her son who worked in one of the wool mills. They were a very religious family and they wouldn't have the radio on. We had to sleep downstairs but weren't allowed to use the toilet which was upstairs. We had to use the one at the bottom of the garden – a plank of wood with two holes in it. My brother and I don't remember any books which is sad because we wanted to learn. They didn't really teach us anything at the parochial school. When we came back to Walton the war was still going on and we had to sleep in the sheds outside because the airmen took up all the rooms. It wasn't a very nice life but you lived with it. In the winter time though we would always have a fire in our lounge on a Sunday which was nice and we would go to see friends to play Monopoly.

BE: *Where were you allowed to go during the war? Was security tight?*

BH: You could only go to up by the Jews' Home and past there was a wire fence. You couldn't go beyond that point. My brother and I remember vividly looking out of our bedroom window one day to see the golf clubhouse up in flames. I think it was a lightning strike. We could of course see the clubhouse from our road.

BE: *Can you talk a little more about your role in sport locally?*

BH: At school I would swim, run cross country, play football and cricket. I played football for Walton before the Frinton County Youth Club and Geoff Ivory, my brother and I would play table tennis at the YMCA on a Saturday morning. We started when we were fifteen. We played in the league for several years as we did in snooker and billiards. When I was in the RAF I would play football and table tennis for them. We gave up table tennis for quite some time when we first had our own families but started again and gave up in about 2008 – unfortunately we found it too quick for us.

BE: *You've been a member of the YMCA all of your life, when did you get involved in leading it?*

BH: John and I didn't get involved seriously until about 1970. I joined in 1947 for about 3 shillings but as I was in the forces and moved away to Colchester etc I wasn't then involved for a while. Between 1965 and 1970 the YMCA itself didn't run. From 1971 I was on the committee. I did 187 meetings over 25 years as secretary, beginning in 1984. The YMCA started in 1919. They had a hut at the back of the current premises. There were 10,000 of those buildings given out by the War Department as they weren't needed. In 1936 J.W. Eagle along with other dignitaries laid the foundation stone. We had about a hundred or so members until a few years ago when it depleted. Various organisations still use the main hall however.

BE: *What are some of the big changes you have noticed in the area?*

BH: Firstly there is the lack of a real community, secondly the lack of locals and all the building work on the outskirts means that you could be anywhere in the country. Sadly you'll never go back to how it was. All the local organisations are slowly disappearing due to both lack of interest and age. People are living a different kind of life now.

BE: *Do you remember the boating lake?*

BH: Certainly. My brother and I were invited to a party at the British Legion. There was a bridge separating the two parts of the boating lake which is still there – one for the toddlers and one for the sailing boats. For some reason my brother and I fell in on the way. There was an area which had been broken, we tried to jump it and missed so we couldn't go to the party. Sailing on the boating lake was wonderful. There were little paddle boats and it's a shame that it is how it is now.

BE: *What are your memories of the '53 flood?*

BH: I was on the Naze at the time and so I was cut off from the rest of Walton because the water went across the Prince's Esplanade. All the huts by the coastguards were washed away. The tide came underneath them, picked them up and took them out like matchboxes. Temporarily the Naze became an island. I was engaged to a girl in Frinton called Mary Broughton at the time and I was at her house the evening before. I said that I should probably get back as the tide would be coming in but when I was back I was stranded on the Naze.

5. Mike Pratt

Mike Pratt was the son of the gamekeeper at Walton Hall and grew up at Creek Cottages. I asked him to describe his memories of Walton Hall Farm.

MP: The Naze was a separate community to Walton and the farm was even more isolated. The general public stopped at the bus stop near Sunny Point and there was very often a locked gate by the farmyard so access was restricted most of the time. My position at the Naze was the son of the gamekeeper, Bill Pratt, who started as a gamekeeper at Marsh Farm, the Ashes and probably Birch Hall until he was 19. My father went into the RAF in 1939 where he was trained as an airframe fitter and he was at Manston for the Battle of Britain where they had fourteen Hurricanes which lasted ten days. He then went to Scotland with two Hurricanes and was then transferred to Canada before finally coming back for the D Day landings. He was in Herefordshire on towing planes and gliders when he met my mother. After the war he went on the railways as a fitter for a bit and in 1950 went to Walton Hall as a gamekeeper, a job which he did for 58 years. He was very skilled and knowledgeable. I can't remember exactly when but there were seventy five brace of English partridge shot there one year. The Walton Hall shoot was very popular and was held in high esteem as a mixture of driven birds and wildfowling on the marshes. Father was also a dab hand at going to get lobsters at low water when he would go along with a hook and a net. I remember one day when I was about 16 he had so many lobsters that he had to put most down and only take the best. He couldn't carry any more. He had an uncanny knack with game and gamekeeping. One day he was watching some cricket indoors, drinking his home brew and everybody else was out by a piece of wheat watching the combine so that when the foxes came out they could shoot the fox. Just before lunch he went out with his 4 10, stood at a particular place and didn't really have to wait. Two cubs ran out which he shot and then he went back to watching the cricket, much to the annoyance of everybody else who'd been there all morning. Another story of him is when everyone was shooting down at the Splash. One of the guests shot some bar geese, which were protected. John Eagle had said that everybody had to eat everything that they shot, knowing that bar geese can't be eaten – they are foul. The guest who had shot the goose said 'how are you supposed to eat this then?' Bill Pratt said 'you put a house brick inside the bar goose and put it in the oven for three hours, then you take it out and eat the house brick.' Another story of him involves the Snow Goose film (1971) some of which was made up at the Naze. Sometime after the filming my father told me 'there's a man who comes up as part of the Snow Goose, bit of a rough looking chap. He stops at the gate and wants to go through' (he got the normal Bill Pratt interrogation). Then father said 'I went to get the key and he followed me in' and they sat and had a drop of home brew. Later in the week the same chap appeared and went straight into the house with a bottle of scotch heading in dad's line. We thought no more about this and then one day we were sat watching a film and Richard Harris was on it. Father said 'that's the man who kept coming to the cottage!' So Richard Harris had worked out that rather go and wait down at the bottom, he'd go to the gamekeeper's cottage where he was very welcome. The gamekeeper had similar interests and they spent two or three good afternoons together.

Personally speaking I spent a lot of time with him. I stacked bales with him on the farm when I worked in the summer, we went shooting rabbits, quite often twice a week, with a tractor with the front windscreen out and the 4 10. I was told off if I couldn't hit the rabbit in the head because they were going to the butchers. I'd go and do a bit of wildfowling with him round the wall. Nearly every Sunday morning he went ferreting. We ran nets on the beach by the pile of rocks for soles in the summer and cod in the autumn. He was always on the go at something and of course he always wanted to be outside. Once he left the farm and moved to High Tree Lane I think the end was near. The farm was him and in actual fact although he didn't own it, it seemed like it was his. John Eagle would have friends who wanted to walk the farm and he used to say 'well you better go and see Bill'. John couldn't give permission for people to go up there without Bill! It was a funny situation but the two of them were great friends.

I was born in 1948 at Marsh Farm. I lived in one of the two cottages which have now burnt down. Nan and Granddad lived up there as did Fred and Juan so it was a bit of an enclave. In 1950 father went to 5 Creek Cottages. 1953 is the first year I truly remember. I had scarlet fever and was isolated for 6 weeks. Nobody could see me except through glass. If you go in 5 Creek Cottages today there are still holes in the door of the small bedroom where they fumigated the room.

The next thing I remember is the flood. I looked out of the back window and saw that water had come all across the farm, flooding the lowlands and getting to the edge of the hill. The next day I walked round and there were lots of dead sheep which had floated across from Horsey Island through the two big holes in the wall. I remember dad burying a cow. He dug a hole alongside the cow and then I believe, with a horse, rolled it over into the hole.

Most of the boats that worked in Walton went from Eagles' Hard. They were mostly like the Bumblebee belonging to Jack Hipkin which was fairly small, about 21 foot and they went out lobstering with hoops. There were a load of lobster boxes in the Twizzle. There were also lots of fishing parties. My father took fishing parties himself in either Jack's boat or Bob Hipkin's boat.

When I grew up the farm was arable and had a dairy herd. The cowmen were a Mr Griggs who lived in the northern cottage of Walton Hall cottages and Mr Reeve lived in the other one. Later on Mr Andrews came to the farm to be the head cowman. When the herd went there were still heifers at the Naze which fed in to the herd at Devereux. On Christmas Days father would get the feed ready on his tractor and trailer.

I'd drive the tractor and he would shovel it off into the manger so as to be done as quickly as possible. Boxing Day was similar so he had a couple of semi-free days. The heifers at the Naze stopped when the herd at Devereux left in 2001. When I was a child there were working horses. I think there were two for the farm and one for the cowmen. They were Suffolk Punch type horses. You often saw them with a tumbril on the back. Sugar beet was pulled by hand with a hook. You'd put a spike on the end, pull it up, chop the top and bottom off and throw it in the tumbril so it was quite labour intensive. The tumbril would be used for muck spreading also. It would be taken to various intervals in the field and then there was a rather interesting implement which went on the back of a tractor with a big spinner which threw the muck everywhere. At harvest time the horse would go round the field with a little trolley behind it and the trailer would be hooked to the trolley with a pin. When the trailer was full of wheat or barley sheaves it would be driven to the stackyard with a tractor. I remember going down to 10 acres one day and father had finished. He was with Bert Wilder. He took the horse out of the shafts, put the traces over the top, put me on top and said 'drive the horse back to the stables boy'. I went back up to the farmyard, round the back of the thatched cart lodge and in to the stables. There was someone there to take me off. They fed them mainly chaff not many oats as they didn't want them to be too frisky. The next day I went down to Walton with mum on my bike and the horse went home on its own.

When I was a child there was quite a lot of evidence of the war at the Naze. For instance there were two aeroplanes. On the marsh there was a Halifax with a big debris trail showing where it had crashed, engines and lots of oxygen bottles. Over on the saltings there was a Hurricane, now quite famous as it's in Hendon Aircraft Museum. It was retrieved by Geoff Rayner and he wrote a book about it. In Thistley Field, opposite the sewage works, there was an army camp. On the southern end of the field there's a pile of rubble which used to be a roadway. On the beach there was still occasional evidence of the tubular tank traps that had been constructed all the way along. About two years ago I picked up two pieces of that on the foreshore and put them in the Heritage museum. All over the farm there were loads of pickets which we took to use to secure our nets on the beach. There are obviously still the pillboxes and when I was a youngster there were piles of mines just over a foot across, by the sewage works, which had been made safe. There was one occasion, when the farm had a better subsoiler or chisel plough which could go a little bit deeper. They had chisel ploughed sheep marsh, then harrowed it, then Arthur Mills had rolled it a couple of times and he noticed a mortar shell. The Naval Bomb Disposal was sent for. My father walked out to the shell with the bomb disposal expert and said he reckoned it was a dud. The man put some explosive down, a piece of fuse and got everyone to the fleet. It went off with an almighty crack and left a hole that you could fit a small car in. The roads across the Links from the farmyard were to take the ammunition to the ack-ack guns. I believe it was that set of guns that shot down the Halifax. The round footings for those guns now lay out on the beach.

We should talk about shooting. Shooting at Walton Hall was a seemingly religious activity. Nothing got in the way of shooting. At its height there were two or three shoot days a year, of some renown. It was a collection of guns at one end of the barn and beaters at the other which prevailed for quite some time. Shooting would be a day of driven birds followed by an evening flight on the splash on the marsh where there was teal. When I was a boy most shoots were half a day at Walton Hall and half a day at the Ashes. Later on it became Walton Hall for a whole day. Lunch would be at Devereux farmhouse before returning to the marsh for the evening flight. There were lots of rabbits shot at the Naze. I was out with father many nights shooting, mainly with a 4 10. Sundays were for ferreting. The farm was a great partridge farm for grey legged partridge which sadly gradually disappeared. The reason is not quite known but probably to do with the introduction of spray which killed insects that grey partridge fed on. There were water voles alongside the fleet next to the sewage works but I haven't seen them for years. Foxes were a constant thing. I have a lovely photo of father holding up two foxes that he shot before breakfast.

I'll say a little bit about where I lived. Creek cottages were built for the coastguards when they took their people out of the watch vessel creek which was in Cormorant Creek. There are tales that the children who were there would walk round the wall to school. Creek cottages were built in 1903. The electricity was connected when I was very young and the flush toilet when I was about nine. Prior to that we only had a closet and a tin bath in front of the fire on a Sunday night if you were lucky. There are five cottages: four for families and one for a bachelor. In number one was Bert Wilder, Mrs Wilder, Janet and Jill Wilder; in two was Ben Knights, Mrs Knights and Joan but Joan had moved on and was married soon after we moved to number five; in number three, which was the smaller house, was Wilf Russ and his mother – I think Wilf's father was a shepherd at the Naze when there were sheep there; in number four were Mr and Mrs Andrews, Sue and Joan; and we were in five. In Hall cottages there was Mr Griggs the head cowman in the north one and Reeves the other cowman in the one nearest the Hall. There were another couple of cottages in High Tree Lane where the Mills lived. The farm was fairly isolated and I didn't know many people who lived on the Naze, let alone Walton. Other people on the farm were Tyler, a horseman and Arthur Mills.

There was loads of sugar beet and acres and acres were chopped, by hand, with a six inch hoe to make them six inches apart. It was priced by the acre and I chopped some when I was a student for some extra cash. My father's thumb on his right hand actually stuck out because of how much he hoed. It was piece work and a chance to make a bit more money so father would often have his dinner and then go back down to the beet in the evening. They were chopped in May and June. Father once had a falling out with Woodruffe (JW Eagle) over the price of sugar beet. There was a farthing mentioned and Woodruffe refused. Three or four days later Woodruffe had a walk around the field and liked what they were doing so he said he would pay an extra ha'penny, very typical of the days.

I'll move on to machinery on the farm. I remember seeing the thrashing machine in the thrashing yard. All the wheat and barley would be taken to the stacking yard where the Dutch barn is still. It was thatched with straw and left there to be thrashed through the winter. The first tractor I can remember is a Fordson petrol/paraffin. In the early days you would have a dedicated tractor driver. Nobody else would be able to touch it – it was a weird and wonderful thing. They would start it on petrol, switch it to paraffin before it was hot and then it would slowly die. The first was replaced by a Fordson Major. I remember seeing the Majors with a binder on the back and a baler on the back with a sledge on the baler for square bales. Then, Fordson 4000s came on the scene. After several years a 5000 came to the farm, thought to be a big tractor at the time but now it would only be seen as a little thing to run down the road on. Tony Strange watched it arrive, he was going to be the driver, and most of the farm management were there to see this brand new tractor. They said 'what do you think of it Tony?' He asked the man who was delivering it 'does it have a radio?' The man said yes. 'Does it have a heater?' and the man said yes. Tony Strange then declared it as a good tractor.

Everyone referred to him as Norfolk because he spoke with a Norfolk accent.'

Since Mike's childhood Walton Hall farm has changed beyond recognition. In our conversation he referred to it as the 'true end of the feudal system', with labour now cut ten-fold, cropping far less diversified and the farm shoot as good as gone. The changes at Walton are a cog in a chain of immense change across the country as agricultural businesses have modified to fit global markets, a transformed labour market and the common agricultural policy.

6. Isobel Ruckert

Isobel's childhood experiences, growing up next door to the Samuel Lewis Convalescent Home, have given her a unique perspective on life at the Naze and the changes that have taken place. I wanted to speak to her to discover more about the Samuel Lewis, a significant feature of the built landscape at the Naze until its demolition in the 1970s, but also about her own childhood at the Naze more broadly.

BE: *Can you begin by explaining your background in relation to the Naze?*

IR: My mother and father, John and Muriel Ruckert, moved to Sunny Point on the Naze, next to the Samuel Lewis Convalescent Home in 1961. I had a very happy childhood. My holidays and life were spent on the cliff and foreshore - playing, crabbing, hunting for shells and shark teeth and particularly enjoying the flora of the Naze. I went to Walton Primary School and then to Grammar School in Clacton. Later, I worked in Walton as a district nurse for some time in the 1980s, covering Walton and the Naze.

BE: *Can you tell me more about the houses on Sunny Point?*

IR: Our house was built in 1934 along with eight others of a similar style by the Franklin family. They had the largest house with the biggest frontage on the end of Sunny Point. Some of the houses were fitted with oak panelling, which I understand was Austrian oak and the floors were English oak. The timbers in the houses were ships beams. During the war the houses were requisitioned for officers and some of them suffered some damage.

BE: *In 1961 when your parents moved to the Naze there was still a single pill box on the top of the cliff. How do you think the Naze has changed since then?*

IR: There has been much coastal erosion which was very frightening for us as a family. In 1974 the cliff directly in front of these houses was protected with an enormous amount of work. Mr Franklin, who lived in Greengable Cottage, owned the road and he sold it to the council for a peppercorn amount so that the works could take place. I remember I was sitting my O levels at the time and because of the cliff work my bed would move around the room. There are still some cracks in our house due to the work but generally we are very grateful that it was undertaken.

BE: *Can you explain your experience with the Samuel Lewis Convalescent Home and the role it played as an institution for the Naze community?*

IR: The Samuel Lewis was built in 1906 by a Jewish board of guardians. Our kitchen and sitting room overlooked the front garden of the building. It was very well populated in the summer from people recovering from surgery from London. The residents would often be sitting in deck chairs and there were two beautiful shelters that people would also sit in. They would usually stay for about two weeks. Changeover day was always a Monday and coaches would come to collect and drop

211

them off. I had a little job to collect all the deck chairs at the end of the day and put them in the shelters so I was able to speak to some of the residents. They were generally very grateful to be having a holiday in Walton but they found it somewhat of a regime. Nursing and hospitals were quite formal then and I gathered from the residents that this was the case at the Samuel Lewis Home. Our leader was matron. She was a permanent member of staff and had a lovely dog – a Dalmation called Spot. Spot was often in the gardens. He was a very gentle dog and the residents would often play with him. As a child all the residents seemed quite elderly and some of them were indeed quite frail. There was a mortuary at the Samuel Lewis and people would occasionally pass away. The nursing staff (and I've met many of them subsequently because of my own work) have very fond memories of working there but apart from matron they were generally temporary staff who would come and do a summer job and then move on once they had found a position in London. The team of staff were very friendly and the building utterly spotless. Local ladies worked there as cleaners and both the house and gardens were kept to a very high standard. There were two wings to the Samuel Lewis – the West and the East – the ladies were accommodated in the east wing and the gentlemen in the west. Although the main entrance and the frontage overlooked the sea, the only entrance that was really used was the one at the rear. The thing I was most fascinated by was the very formal clothing of the residents, bearing in mind they were on holiday but that was a time when things were just beginning to change from when people used to go to the seaside in their working clothes – they would roll their trousers up and wear a flat cap. The residents had very formal meal times and break times and I they said to me that they would have liked more cups of tea in the afternoon and coffees in the morning rather than the regimented 'tea times'. I remember a beautiful mosaic in the entrance hall which was unfortunately destroyed when the building was demolished – it was more sacked than demolished. In the end the workmen became fed up and set fire to it to speed the process up which was quite terrifying.

BE: *Can you talk in more detail about that 'end period'. Why was the decision taken to close the home?*

IR: The concern was two-fold. Firstly, there was concern about the erosion of the cliff and there was also a change of thinking as regards to convalescent care that perhaps people shouldn't be moved out of London on a bus and then taken down to Walton. It was very cold here in the winter and the Samuel Lewis was open all the year round. However, the major concern was the erosion. The Samuel Lewis had two beautiful gardens. It had one more formal garden which was confined by the walls. Then there was the little road with metal railings and a gate. In front of that was a very beautiful informal garden which slid down to the sea. I can remember residents sitting down there on deck chairs. There are now garden flowers growing all over the cliff down from Sunny Point and they are flowers I remember having been in the 'informal' garden – red hot pokers are the main thing I noticed.

When things stopped it all happened very quickly indeed. The residents were moved out and the building left standing but it was abandoned. The garden became completely overgrown before it was demolished following the work on the cliffs in 1974. There was great sense of sadness in the community that we would be losing this building and also it was a source of employment for people in Walton.

BE: *What of religious observation?*

IR: That was very strong indeed. Passover was observed and the Rabbi would come to stay in our house, staying only for the duration of Passover and other religious festivals to observe them in the Samuel Lewis. He brought kitchen equipment which was only used once a year but then the food would be kosher but he was quite a fascinating man with his shawl and his prayers. He would always pray before his meals.

BE: *Was there any form of public backlash towards the closing of the building?*

IR: Yes. Initially people wanted it to be preserved. Local people really didn't want to lose this beautiful building. They would have preferred for the building to be converted to flats instead. The difficulty was that the garden was so enormous that it was a developer's opportunity. Initially there was a plan to demolish the building and build enormous flats. There was much public uproar about this and the plans were changed to have bungalows at the front and houses at the back.

BE: *You went to Walton Primary when you were growing up. Several people have mentioned to me about the perceived division between children at the town end of Walton and at the Naze end. Is this something that you experienced?*

IR: Yes. We children at the Naze were quite close friends, mainly because of the journey to school. In the winter, for a few weeks we were allowed to travel home on the bus. My closest friends came from Second Avenue which was a council estate with homes for young families. We organised our own games. We would play football with the boys as there often weren't enough of them to make a team. We would organise fetes with the guidance of adults to raise money for various charities. We also organised jumble sales and manned the sales. We spent a lot of time on the cliff. The boys would often play wars – the pill boxes, gun mountings and barbed wire provided for a great sense of fun for small boys. The Baptist Church was very lively and as children we were made very welcome there. We also had a 'friend for a fortnight' when the families came down in the summers with their caravans. We would go to the beach, particularly Robert Hipkin's beach, taking our crabbing equipment with us. Some of the families would invite me to the pier on the Friday night before they were going home and my mother would give me pennies to spend on the pier. I didn't generally go on to the caravan sites though. There was a general perception that these people were different to us. They were mostly Londoners and there was an issue of child safety. You didn't go there and didn't wander about near there. There was also the Childrens' Special Service Mission at Robert Hipkin's beach and the Albion Beach. The young people used to stay in the Mabel Greville and we would go to the beach and have a sing song. I can remember a piano accordion worked with the feet. We all sang the choruses. Behold I stand at the door and knock was a particular favourite. They would also do little campsite suppers. So, it wasn't all focused on religion and Christianity. It was about having summer fun as well. There was a great simplicity about it all and lots of local Naze children would join in.

School was a very happy time. Miss Prosser was out Headmistress and was a redoubtable Welsh lady. I also particularly remember Mrs Betty Price who was a stickler. We all lived in fear of going in to junior one because that's when you had to learn your times tables. She was however largely a fantastic teacher. She did once tap me on the head with a ruler which I had previously hurt when I fell out of a tree and banged it. Where she tapped me I passed out. Poor Mrs Price was horrified and my mother had to be summoned.

BE: *What were your experiences of the backwaters?*

IR: My father had a boat, a little rowing boat first of all and then he bought this cabin cruiser and would go far out to sea fishing. I remember when we had only been here a very little while, I think it was the first summer, my father decided to take me to Stone Point but we were caught by the tide. I remember being put on his shoulders and the water coming up to his chest and him wading through the water to get back to the Naze. We were absolutely soaked and my mother, who was usually a very mild mannered lady, went ballistic and said 'don't you ever do that with my child again!' Sadly my father died in an accident at the Old Hard when I was just thirteen so we have very sad memories as a result but local people were wonderful at the time.

The other thing I remember are the shops. We had the Blue House dairies which was a wonderful shop. There was a butchers, Lilleys the baker, a hairdresser, the Foundry was active. My father and I would enjoy walking right the way round the Naze. He would encourage me with a piece of chocolate on every bend!

I have had brilliant experiences of Walton and it has always been a genuinely friendly place. There is certainly a sense today, especially through the Churches, of more help being available for those in need. I remember a feeling of warmth and kindness from the community from when I was a little girl growing up, to when my father died in 1971 and that feeling is still around today.

7. David Eagle

Finally, I spoke to my father, David, about his own experiences of the Naze and Walton Hall Farm as well as his thoughts for the future.

DE: I was born in 1956 into the family that's farmed on the Naze since the 1880s. The family has farmed at Devereux Farm in Kirby-le-Soken and on the Naze, the Naze farm being based at Walton Hall since the 1880s. My father was born and lived at Walton Hall farmhouse until he was seven, and the farm buildings at that stage were fully utilised with a dairy, a brewery, stables for heavy horses, a small granary, and all the accoutrements that went towards it being a self-sustainable farmstead. In the Fifties father still farmed what is now the public open space. Today, as well as the farmland to the west of the open space the farm still includes three hundred and fifty acres outside the sea wall including Stone Point, Stone marshes and, untypically compared to many places, all the foreshore around the Naze, which includes the beach running along the Naze cliffs from an area of fifty five feet from the base of the cliffs down to the tidal mark.

BE: *How has Walton Hall changed through your lifetime?*

DE: I can remember cattle grazing on the public open space. It was used for young stock but at that stage (1960s) the field structure had gone. The farm dairy moved away from Walton Hall to Devereux Farm. I cannot remember, bearing in mind I was born in 1956, seeing cows being milked at Walton Hall but it was a very traditional, line milking system. In my lifetime Walton was the place where young stock was kept during the winter. The young stock included calves that had been weaned, so those three months and older. The farm buildings in the '60s still retained the old thatched stables at the front, opposite the pond. That fell in to serious disrepair though and was demolished. The rest were brick and tile roofed. At that particular stage all the buildings at Kirby were modernised to take the two dairy herds as there were then. So, Walton Hall farm buildings fell into redundancy. There was a lot of straw stored there in small bales and latterly in large bales. There was a big maize silage clamp to feed the youngsters. The staff there at that stage included Bill Pratt, who was the gamekeeper and lived in number five Creek Cottages, Freddie, his brother, who looked after the young stock, Horace Mann, who lived in 186, Elm Tree Avenue, and Arthur Mills, who lived in Walton. The Pratt family and the Mann family had worked for the farm for two or three generations. Most members of those families had connections with the farm that went back to their childhood.

BE: *What about changes to the physical landscape itself?*

DE: When the land was sold to the Council in 1963, at the Council meeting that decided upon the purchase there was intent not to maintain breakwaters in the future. By the 1970s the Tamarisk Wall, which had breached in the 1953 flood, was in a similar state to today. I remember in the seventies, the Council graded the cliffs in front of the Samuel Lewis Convalescent Home, now a bungalow estate, which was a fairly radical move but it was part of a Council investment to stabilise the whole of the coastline, bearing in mind there were a lot of beach huts on the corner by Hipkins. The two public inquiries that happened on the Naze were in the fifties which I obviously have no recollection of but clearly, at that time, there was substantial intent to develop the Naze, because those public inquiries were talking about hotels and something like 350 dwellings. Interestingly, on the western side going along the private road [Old Hall Lane] leading down to Walton Hall there is an area of land which was then highlighted for an underground car park, which suggested that there was sensitivity in the planning process towards the view from the Naze across to Hamford Water.

BE: *How reliant do you think Hamford Water is on the Naze as a landmass?*

DE: Hamford Water is a two thousand hectare internationally protected wildlife sanctuary. If you can consider a tidal area being a protection zone the North Sea starts on the frontage of the Naze cliffs. Between the Naze and Harwich we have Pennyhole Bay which has a shallow area of shifting sands and mud. I would consider that to be one of the first lines of defence against the North Sea because it breaks the wave energy. Clearly the Naze is the primary structure that protects the whole of Hamford Water. We used to see one and a half metres of erosion along the frontage but with climate change this is increasing and we are seeing increased rates of erosion along different areas of that frontage. That will impact in the longer term if nothing is done. Any action will contravene the aspirations of allowing the cliffs to erode to expose geological structures, which creates a very difficult political issue. We have to consider that unless one is prepared to build and maintain (and that's the issue – you can build a concrete frontage but maintaining that frontage is going to be really difficult in times of climate change) concrete frontages, the landscape will erode. That means that, within a century, the North Sea will be, I believe, banging on the door of Hamford Water. It is unlikely that Hamford Water will be the same as it is now. The amount of silt coming into Pennyhole Bay from various sources is silting up the Wade and the back reaches of Hamford Water. So, if one considers that as the Naze erodes, Hamford Water is growing, one could say that Hamford Water is reverting to a landscape of more salt marsh and that as the land rises the amount of water in Hamford Water decreases. The ability of land to revert to terrestrial land rather than tidal land will, I believe, become a real possibility in certain areas. The only reason that might change is through human intervention, dredging, engineering operations, in order to maintain the yachting industry which is very important to Walton's economy. So, it's a mixed bag. On top of that there are the issues with Harwich Harbour and their aspirations to maintaining deep water channels as well as at Felixstowe and how that will impact upon Hamford Water in terms of changing the amount of sediment in the water and the beaches on Stone Marsh, Foulton Hall on the other side and north Horsey.

BE: *There has been a farm on the Naze for more than 1000 years. How important do you feel it is and how viable do you think it is for there to be a farm on the Naze in the future?*

DE: Where there is land and where there is land in private ownership land will need to find a purpose. The purpose you put that land to depends on the viability of the enterprise you are looking to undertake. Agriculture is the norm if there is enough acreage to make it worthwhile. The Naze farm is only 350 acres of prime land, which has a capacity to produce 3.5-4 tonnes of wheat per acre, which is still a significant amount. Unfortunately 350 acres is not enough in a viable farming context. You need about 1500 acres to make a viable straight forward, commodity growing, arable unit. Therefore, Walton Hall has to be attached to another farm somewhere else, if you're going to use it in that way. The downside is bringing big machinery through the town and the practicalities of that. It is an out station as a farm so it's potential for conventional farming is limited. In two hundred years' time we have to question what the landscape will look like. If someone has the foresight to make sea walls that maintain strong defences it probably means that the sea walls will look very different. They will go right back to the high ground and I think the slope on the back of the sea wall will become an extension of the marshes and, quite possibly, if you were cropping it you would crop to the top of the sea wall. Alternatively, you could be looking at grazing there. However, grazing comes with a lot of issues in terms of management and how you square the economics. It would have to be a big enough unit so, if you were to keep say 200-300 beef there, maybe 800-1000 sheep you would need to invest in the buildings and in so doing you would be making a sizeable presence there for the future. That could happen if there is the will to do so, but you have to also consider what the state of Walton will be in two hundred years' time. Will the local Council be able to protect the road running along Prince's Esplanade indefinitely? It would certainly become a very different town if it were to breach there. The Naze would become a satellite of Walton as opposed to being a joint part of it. Maybe if the Naze was an island connected by a bridge (or causeway) it would give it a new lease of life. The other thing to bear in mind is the sand that protects the frontage. In great grandfather's time the farm needed to put two teams of horses on grain wagons in order to pull grain from Walton Hall farm through to Brookes of Mistley and that meant pulling it through the sand dunes. That has now long gone. The sand on the frontage will only be there if the breakwaters are protected.

BE: *What are your memories of Walton Hall house when you were growing up?*

DE: In the sixties, one could relatively safely go up the stairs, through Walton Hall house and go up on to the roof. I remember that, having gone out on to the roof, you could look through one point and see through all the floors down to the cellar. It was however certainly safe to go up there. In terms of some of the architectural features within it, coming down from the roof staircase there was a little area of circular doors called 'Piccadilly Circus' which looked in to all the bedrooms in that area. By its very nature though it was enclosed and there wasn't any natural light in the centre of the house. The house has always been known as very dark and any house that is dark seems to conjure all sorts of strange folklore. Father used to say that, as a house in which to live, it was bitterly cold in winter. It had a lot of fireplaces but in winter everyone tended to huddle around one. There was a huge cellar and in grandfather and great grandfather's time there was a lot of very fine wine down there. When we sold the house in 2001, there was a range of buildings that ran from the main block, going west, which included the kitchen, the game larder, a little dairy and a larder at the end and outside there was a huge, brick, enclosed pit with a man hole on top that must have gone down twenty feet which I assume was originally put there for collecting roof water – although having said that, wine bottles were at the bottom of it. Father also said that during the war there was an Indian regiment based at the house and they burnt anything and everything because of the cold winters, especially in 1940/41. Also during the war the cellar, which had been bricked up in order to secure the contents was broken in to and the contents disappeared. There was a second story window on the north side which, when the building was being redeveloped, was stripped back and opened up. It seemed as if, above the window, there was the slotting mechanism for a portcullis, suggesting the block house's purpose as a building to defend against sea robbers. One would conjecture that that window was the main doorway in to the building against which you would put a ladder to gain access to the main block. There wouldn't have been any windows downstairs.

The dining room has always seemed like a good, functional room. The lounge, a big square room, had two large plaster recesses on the north side of it. I don't know what the fireplace was like because it wasn't there by the sixties. A lot of features had disappeared, including the lead off the roof, the fireplaces and the doors. All the windows were shuttered and had quite deep recesses. If one goes back to that original 1560s concept of it being a defensive property, one can understand why the bottom parts of the walls are so thick. This is however sadly all conjecture. The kitchen was quite small but, as you go down to the outbuildings on the north side, the bottom one, which was attached to the red brick wall that ran down through the garden going east west, had a lot of slate shelving in. All those buildings however had lost their roofs by the 1980s.

BE: *In your opinion what would a sustainable future for the Naze look like?*

DE: On an environmental level I believe that one of the key habitats in Hamford Water is fresh water. The emphasis in recent years has been to create inter-tidal areas through managed realignment. Personally I believe that if the Naze went for managed realignment one would downgrade the emphasis on protecting the Naze as an integral landscape. I believe it is essential that you maintain the sea wall, that you make a stand as to where the sea goes and that you maximise the enterprises to give a reason to maintain those sea walls as long as you possibly can. The farm needs to find its new identity as a small, self-contained unit. It could be an arable farm but on the basis of processing on site and adding value to a commodity, making employment viable. If you create employment viability on the Naze then you create something that really has a future. The alternative is that you let the Naze go to sea. In the next fifty years a choice must be made.

Acknowledgements

This project would not have been possible without the support and enthusiasm of numerous people. Thank you firstly to Dr Chris Gibson, for writing the foreword for the book but also for his ongoing support and comments on various drafts of the manuscript. His support, as someone who has worked for so long studying and protecting the Essex coast has been invaluable. Thank you also to my brother, Sam, who designed the cover of the book and my sister Emma who was particularly helpful during the design stage.

I have been very lucky to speak to many people who have had life-long experiences of Walton and the Naze, its special landscape and unique wildlife and geology. Several of these people have been willing to be interviewed and I have transcribed these interviews in the appendix of the book. They include my father David, Mike Cranstone-Todd, John Fleming, John Halls MBE, Bryan Hicks, Mike Pratt and Isobel Ruckert. Thank you also to Bill Woollard for speaking to me about his memories of the Maunsell Sea Forts and to my grandmother, Daphne, for telling me about farming in the local area in the 1950s and 60s.

In terms of archival research I owe a great debt to John Button who has been involved in the project throughout and has provided me with access to the Richard Stone diaries which he has transcribed, as well as translating the 'Domesday of St Pauls' document. Without John's work, enlightenment of the medieval history of Walton would have been much more difficult. I have sometimes used John's work literally and want to acknowledge the extent of his work on the medieval history of Walton. Many thanks also to Liz Bruce, archivist of the

Frinton and Walton Heritage Trust, for her enthusiasm for the project and willingness to help which has been invaluable. I have spent several very happy and interesting mornings at the FWHT archive, located at the Walton Maritime Museum. Thank you to the Frinton and Walton Heritage Trust for allowing me access to their superb archive. My thanks also to the staff at the Essex Records Office, an archive where I was able to illuminate on much of the nineteenth century of Walton Hall in particular, and to the staff at Frinton, Walton and Colchester libraries. The local history collections at these libraries have been very helpful in giving me context. Thank you to all the historians who have previously researched and written about the history of Essex and Walton, particularly Dr Peter Boyden and the Victoria County History group.

Thank you to Essex Wildlife Trust for giving me the opportunity to work at the Naze during the latter period of writing the book, a period which has been highly valuable in broadening my perspectives. Thank you especially to all the Essex Wildlife Trust Naze volunteers – especially Sue, Gerald, Eliott, Wendy, Stephen, Alex, Rose, Isobel, Mike, Emma, Ginny, Peter and Ben for continuing to inspire me and for your passion for the Naze. I am excited that Essex Wildlife Trust will have an institutional presence at the Naze from 2016 with the opening of the Education and Visitor Centre and I look forward to seeing the centre's influence develop over the coming years. Thanks to all at the Naze Tower, especially Michelle, Ted and Rosemary for the information relating to the Tower and for the regular mugs of tea!

Thank you very much to all who have read through and commented on the various drafts of the book including Dr Ian Wei, Professor Ronald Hutton, Dr Andy Flack, Dr Chris Gibson, Zoe Lynes and Laurel Hart. Each of your comments has helped shape the final manuscript for which I am very grateful.

As my first book, the contents have developed as my own intellectual thinking has developed. Key in this process have been my tutors at Bristol University, especially Professor Peter Coates, Dr Andy Flack and Professor Brendan Smith and I want to thank the history department at Bristol for being so supportive and inspirational throughout my time in the west country. I should also mention my history teachers at Ipswich School – Kathryn Galbraith, Peter Gray and Stephen Blunden - who had a profound influence on me earlier on in life, inspiring my passion for the past.

I must thank all of the photographers who have allowed me to use their images, particularly Putmans Photographers who have allowed me to use images from their Walton archive. Finally, thank you to my family for supporting and encouraging me throughout the project. They share my passion for the Naze and I will miss the kitchen table discussions when, invariably, the Naze will be raised as a topic of conversation, which have been plentiful during the process of writing this book.

About the Author

Ben Eagle grew up just a few miles away from the Naze. His family have deep farming roots in the local area and own the farm at Walton Hall as well as land in Kirby-le-Soken. Ben read history at Bristol University, specialising in environmental and animal history, and agriculture at the Royal Agricultural University, Cirencester. He is an Associate of Trinity College, London. This is his first book.

Bibliography

Books

Benham, W.G., *Essex Sokens and other parishes in the Tendring Hundred* (Chelmsford, 1928).

Bloom, U., *The House that Died Alone* (London, 1964).

Boyden, P.B., *The first 124,999,061 years of Walton: A Short Account of the town to AD 939* (Walton-on-the-Naze, 1979).

Boyden, P.B., *History of Walton: Part 2, 1553-1800, The Birth of a Resort* (Walton-on-the-Naze, 1972).

Broadhurst, R., *Managing Environments for Leisure and Recreation* (London, 2001).

Burns, A., D. Keene and A. Saint, *St Paul's, the Cathedral Church of London, 604-2004* (New Haven, 2004).

Butler, A., *Sheep* (Winchester, 2006).

Button, J., Walton in 1297 in the time of Alice Horlefrensche (Lulu Online Publishing, 2009).

Debenham, J., and A. Summers, *The Essex Hundred Histories* (Southend, 2008).

Defoe, D., *A Tour Through the Whole Island of Great Britain: Divided Into Circuits or Journies* (London, 1762).

Doody, J.P., *Coastal Conservation and Management: An Ecological Perspective* (AH Dordrecht, 2001).

Dunnett, R., *The Trinovantes* (London, 1975).

Edwards, A.C., *A History of Essex* (London, 1958).

Edwards, P., *Farming: Sources for Local Historians* (London, 1991).

Fairhall, D., *East Anglian Shores* (London, 1988).

Ford, P., *Tendring Peninsula: land of Milk and Hunnye* (Bristol, 1988).

Gallyon, M., *The Early Church in Eastern England* (Lavenham, 1973).

Garmonsway, G.N. (trans)., *The Anglo-Saxon Chronicle* (London, 1972,1982).

George, W and S. Vincent, 'Report of Field Meeting to Walton-on-the-Naze and Wrabness, Essex, 2.X, 1976 with Notes on the London Clay of Walton.' (2nd June 1977).

Gore, A., *An Inconvenient Truth* (London, 2006).

Grieve, H., *The Great Tide* (Chelmsford, 1959).

Hale, G.H., *The Domesday of St Paul's of the Year 1222* (London, 1858).

Hendy, P.M., *Treacherous Tides* (St Osyth, 2007).

Ingram, J. (trans), *The Anglo-Saxon Chronicle* (London, 1912).

Jacobs, N., *Frinton and Walton: A Pictorial History* (Chichester, 1995).

Jarvis, *Essex: A County History* (Newbury, 1993).

Johnson, I., *Turning Point: The story of Kirby-le-Soken, Essex* (London, 1982).

Kemble, J., *Prehistoric and Roman Essex* (Stroud, 2001).

Lamb, H.H., *Climate, History and the Modern World* (London, 1982).

Martin, G.H. and A. Williams (eds.), *Domesday Book: A Complete Translation* (London, 2003).

Modlen, L.C., *The Story of Walton-le-Soken* (Walton-on-the-Naze, 1955).

Morant, P., *The History and Antiquities of the County of Essex* (London, 1768).

Nicholls, R.J. and S. Rupp, *Managed Realignment of Coastal Flood Defences: A Comparison between England and Germany* (London, University of Middlesex, 2002). available at http://www.survas.mdx.ac.uk/public11.htm

Norden, J., 1594, *Speculi Britanniae Pas: The Description of Essex* (London, Camden Society, 1840).

Norman, B.J., *Walton-on-the-Naze in old picture postcards* (Zaltbommel, 1984).

O'Neil, R., *The Naze Tower at Walton: The First Line of Defence* (Bognor Regis, 2006).

Page, W., *A History of the County of London: Volume 1: London within the Bars, Westminster and Southwark* (London, 1909).

Palmer, K., *Setting the Record Straight: A Concise history of Frinton, Great Holland, Kirby and Walton* (Frinton-on-Sea, 1994).

Palmer, K., *"Wish you were here" in Walton, Frinton and Kirby* (Romford, Ian Henry Publications, 1992).

Paine, D. (ed), *The History and Work of the Walton and Frinton Lifeboat* (Walton, 1975).

Pratt, R., *Aelduluesnasa: Kirby in History* (Kirby-le-Soken, 1976).

Rackham, O., *The History of the Countryside* (London, 1986).

Ransome, A.M., *Secret Water* (London, 1939).

Rayner, G., *Seaside Front Line* (Medlesham, 1996).

Rigby, S.H., 'Social Structure and economic change in late medieval England' in R. Horrox and W.M. Ormrod, *A Social History of England 1200-1500* (Cambridge, 2006).

Ruddiman, W.F., *Earth's Climate, Past and Future* (New York, 2001).

Sawyer, P., *The Age of the Vikings* (London, 1962).

Senter, *A Study of the Lords of the Manor of Walton-le-Soken* (Colchester, 2001).

Steer, J., Walton and Frinton Lifeboat: A Station History 1884-2005 (Lavenham, 2005).

Stratton, J.M., Agricultural Records A.D 220-1968 (London, 1969).

Wade-Martins, S., *Know the Landscape: Farms and Fields* (London, 1995).

Warren, S.H., 'Pre-history in Essex', *The Journal of the Essex Field Club*, vol. 5, (1918).

Warren, S.H., 'On a prehistoric interment near Walton-on-Naze'. *Essex Naturalist*, vol.16. (1911), 198-208.

Weninger, B. et al., 'The Catastrophic Final Flooding of Doggerland by the Storegga Slide Tsunami' in *Documenta Praehistorica,* vol.35 (2008).

Wilkinson, T.J., 'Environment in South Essex' in *East Anglian Archeology,* vol. 42 (1988).

Wilmshurst, T., *A Descriptive Account of Walton-on-the-Naze* (London, 1860).

Wilson, D., *Anglo-Saxon Paganism* (London, 1992).

Wood, E.A., *A History of Thorpe-le-Soken to the year 1890* (Thorpe-le-Soken, 1975).

Wood, E.A., *The End of the Soken Court* (St Leonards-on-Sea, 1954).

Wright, T., *The Picturesque Beauties of Great Britain: A Series of Views from Original Drawings* (London, 1834).

Yearsley, I., *Islands of Essex* (Romford, Ian Henry Publications Ltd, 1994).

Yelloly-Watson, *The Tendring Hundred in the Olden Time* (Colchester, 1877).

Young, A., *General View of the Agriculture of the County of Essex Vol II* (London, 1807).

Sources from edited collections

Hicks, C., 'Annals of the Parish of Great Holland by an Old Stager' in A.F.J. Brown (ed.), *Essex People 1750-1900* (Chelmsford, Essex County Council, 1972), 71-79.

'The diary of Ralph Josselin, 1616–1683' in MacFarlane, A. (ed.), *British Academy: Records of Social and Economic History*, ser. 3 (1976).

Journal Articles

Brooke, C.N.L., 'The Composition of the Chapter of St Paul's 1086 – 1163' in *The Cambridge Historical Journal*, vol. 10, No. 2, (1951).

Cole, W., 'The Development of Architecture in Essex' in *The Essex Naturalist*, vol. 8, (1894).

Coles, B.J. 'Doggerland: a speculative survey' in *Proceedings of the Prehistoric Society* vol. 64, (1998), 45-81.

Dalton, W.H., 'Walton and Frinton in 1902' in *Essex Naturalist* vol.12 (1902), 217-221.

Essex Recusant Society, 'Darcy and Savage' in *the Essex Recusant,* vol.27 (1985), 15-19.

Lawrence, E.N., 'Floods of the Past' in *Weather,* vol. 8, No. 3, (March 1953).

Warren, S.H., 'Pre-history in Essex' in *The Journal of the Essex Field Club,* vol. 5, (1918).

Websites

http://www.britishhistory.ac.uk/report.aspx?compid=74279&st rquery=bessie+Savage

www.british-history.ac.uk/report.aspx?compid=35353 'Secular canons': Cathedral of St. Paul', *A History of the County of London: Volume 1: London within the Bars, Westminster and Southwark* (1909), 409-433. (accessed on 14/07/2010).

IPCC, Climate Change 2014: Impacts, Adaptation and Vulnerability (Fifth Assessment Report, 2014), 368-372 available at http://www.ipcc.ch/pdf/assessment-report/ar5/wg2/WGIIAR5-Chap5_FINAL.pdf

http://www.battleofmaldon.org.uk/ (accessed on 02/02/2012)

http://www.blupete.com/Literature/Poetry/MasefieldSeaFever. htm (accessed on 24/05/2010)

http://www.britishhistory.ac.uk/report.aspx?compid=74279&st rquery=bessie+Savage (accessed on 15/07/2010)

http://www.britishhistory.ac.uk/report.aspx?compid=63925&st rquery=Goldmer+gat (accessed on 1/10/2010)

www.british-history.ac.uk/report.aspx?compid=35353 'Secular canons': Cathedral of St. Paul', A History of the County of

London: Volume 1: London within the Bars, Westminster and Southwark (1909), pp. 409-433. (accessed on 14/07/2010)

http://www.historyhouse.co.uk/placeW/essexw05a.html (accessed on 15/07/2010)

http://www.historylearningsite.co.uk/dissolution_monasteries. htm (accessed on 25/07/2010)

http://hubpages.com/hub/Coastal-Erosion-in-East-Anglia (accessed on 25/07/2010)

http://www.marinet.org.uk/mad/disappearingcoastline.html (accessed on 26/07/2010 and 01/10/2010)

www.nazetower.co.uk/tower_museum (accessed on 12/07/2010)

http://www.nazeprotectionsociety.org.uk/cragwalk.htm (accessed on 15/07/2010)

http://www.nazeman.fsnet.co.uk/page8.html (accessed on 16/07/2010)

http://www.oxforddnb.com/view/article/69349 (accessed 8/01/2012)

http://www.oxforddnb.com/view/article/24723?docPos=1 (accessed on 6/01/2012) Oxford Dictionary of National Biography. Article by John B. Hattendorf.

http://www.oxforddnb.com/view/article/30313/30310?docPos=1 (accessed on 11/01/2012) Oxford Dictionary of national Biography. Article by Stuart Handley.

http://www.oxforddnb.com/view/article/70579?docPos=4 (accessed on 10/01/2012) Oxford Dictionary of National Biography. Article by David Loades.

http://www.oxforddnb.com/view/article/30312?docPos=2 (accessed on 10/01/2012) Oxford Dictionary of National Biography. Article by Geoffrey W. Rice.

http://www.oxforddnb.com/view/article/23647?docPos=1 (accessed on 10/01/2012) Oxford Dictionary of National Biography. Article by Roland Thorne.

http://www.oxforddnb.com/view/article/69349 (accessed on 09/01/2012) Oxford Dictionary of National Biography. Article by John Walter.

http://www.poemhunter.com/poem/sea-fever/ (accessed on 12/08/2010)

http://www.runcornhistsoc.org.uk/rockandclifton.html (accessed on 14/07/2010)

http://www.stosyth.gov.uk/default.asp?calltype=jun03elizabeth one (accessed on 14/07/2010)

http://www.stosyth.gov.uk/default.asp?calltype=jul04medieval priory (accessed on 16/08/2010)

http://www.stosythpriory.co.uk/ (accessed on 06/09/2010)

http://www.tudorplace.com.ar/Bios/ThomasDarcy(1BChiche).htm (accessed on 14/07/2010)

http://www.walton-on-the-naze.com/Nature/nature.htm#The Naze (accessed on 16/07/2010)

Primary Sources, Private Sources, Unpublished Works and Collections

Sources held at Frinton and Walton Heritage Trust archive at Walton Maritime Museum. Courtesy of Frinton and Walton Heritage Trust.

The Diaries of Richard Stone esq., 1810-1835, courtesy of the owners of the diaries and transcribed by John Button BA

Estate maps and records of J.W. and F.D. Eagle farms provided courtesy of members of the Eagle family

Pamphlets and booklets from the Naze Protection Society (NPS) located at Frinton Town Library and Walton library.

'The Walton Report' – put together as a result of the 2003 Countryside Exchange programme that took place between the UK and North America. It was organised by CEI Associates Ltd. Under contract to the Countryside Agency and answering to a National Steering Group. The body was made up from partners from the Countryside Agency, the Countryside Council for Wales, DEFRA, English Heritage, English Nature (now Natural England), the Forestry Commission, the Heritage Lottery Fund, Scottish Natural heritage and the Welsh Development Agency. Members of the BTCV and National Trust were observers.

Essex Records Office sources

- A5899
- D/DU 5/84
- D/DU 408/10
- D/DU 493/12
- D/DU 526/1
- D/DU 954
- D/DU 1444/1
- D/DU 5184
- D/DB M160
- D/DBm M578
- D/DHw E9
- D/DHw E7
- D/DMh/T40

- D/APsW 1131
- D/DFI E1
- D/DAc 239
- D/DAc 242
- D/DAc 243
  D/DAc 235
- D/DAc 236
- D/DAc 374
- D/DHw T14
- D/DCr T5
- D/DCr T1
- D/DCr T2
- D/DCr T3
- Various maps of the county of Essex by Messrs Chapman and Andre (1777), Norden (1594), Speed (1610 & 1676), Jansson (1636 & 1646), Blaeu (1645), Warburton (1726), Bowen (1749) and Gibson and Bowen (1762), located at the Essex Record Office, Chelmsford.

Personal Interviews

Mike Pratt (15.10.14) ; Isobel Ruckert (19.06.15) ; John Halls MBE (30.10.14) ; Bryan Hicks (14.10.14) ; David Eagle (31.08.14) ; John Fleming (27.09.14) ; Mike Cranstone-Todd (26.09.14) ; Daphne Eagle (2.09.14) ; Michelle and Rosemary Nye-Browne (6.7.15) ; Bill Woollard (21.10.14)

Endnotes

1 O. Rackham, *The History of the Countryside* (London, 1986), 29-30.

2 B.J. Coles, 'Doggerland: a speculative survey', Proceedings of the Prehistoric Society, vol.64, (1998), 45-81.

3 B. Weninger et al., 'The Catastrophic Final Flooding of Doggerland by the Storegga Slide Tsunami' in *Documenta Praehistorica,* vol.35 (2008).

4 Jarvis, *Essex: A County History* (Newbury, 1993), 18.

5 S.H. Warren, 'Pre-history in Essex', *The Journal of the Essex Field Club*, vol. 5, (1918), 16.

6 Warren, 'On a prehistoric interment near Walton-on-Naze'. *Essex Naturalist*, vol.16. (1911), 198-208.

7 Warren, 'Pre-history in Essex', *The Journal of the Essex Field Club*, vol. 5. (1918), 9.

8 Warren, 'Pre-history in Essex', *The Journal of the Essex Field Club*, vol. 5. (1918), 9.

9 T.J. Wilkinson, 'Environment in South Essex' in *East Anglian Archeology,* vol. 42 (1988), 101.

10 J. Kemble, *Prehistoric and Roman Essex* (Stroud, 2001), 56.

11 Kemble, *Prehistoric and Roman Essex*, 47.

12 Jarvis, *Essex: A County History*, 19.

13 Jarvis, *Essex: A County History*, 20. Lion Point formed the parish boundary between St Osyth and Great Clacton.

14 Kemble, *Prehistoric and Roman Essex*, 49.

15 Kemble, *Prehistoric and Roman Essex*, 21.

16 Kemble, *Prehistoric and Roman Essex*, 68.

17 Kemble, *Prehistoric and Roman Essex*, 68.

18 Kemble, *Prehistoric and Roman Essex*, 68.

19 A.C. Edwards., *A History of Essex* (London, 1958), 6.

20 R. Dunnett, *The Trinovantes* (London, 1975), 29.

21 S. Wade-Martins., *Know the Landscape: Farms and Fields* (London, 1995), 13.

22 Jarvis, *Essex: A County History*, 25.

23 Edwards, *A History of Essex*, 7.

24 Dunnett, *The Trinovantes*, 30.

25 P. Ford., *Tendring Peninsula: land of Milk and Hunnye* (Bristol, 1988), 16.

26 L.C. Modlen, *The Story of Walton-le-Soken* (Walton-on-the-Naze, 1955), 2.

27 Kemble, *Prehistoric and Roman Essex*, 78-79.

28 Jarvis, *Essex: A County History*, 37.

29 Jarvis, *Essex: A County History*, 39.

30 Not a definitive list of Essex 'kings'. Sourced from Jarvis, *Essex: A County History*, 39.

31 Jarvis, *Essex: A County History*, 40.

32 K. Palmer, *Setting the Record Straight: A Concise history of Frinton, Great Holland, Kirby and Walton* (Frinton-on-Sea, 1994), 11.

33 N. Jacobs, *Frinton and Walton: A Pictorial History* (Chichester, 1995), 4.

34 Palmer, *Setting the Record Straight*, 11.

35 W.G. Benham, *Essex Sokens and other parishes in the Tendring Hundred* (Chelmsford, 1928), 17.

36 Benham, *Essex Sokens*, 17.

37 P. Morant, *The History and Antiquities of the County of Essex* (London, 1768).

38 P. B. Boyden, *The first 124,999,061 years of Walton: A Short Account of the town to AD 939* (Walton-on-the-Naze, 1979), 7.

39 Palmer, *Setting the Record Straight*, 12.

40 Benham, *Essex Sokens and other parishes in the Tendring Hundred*, 18.

41 Palmer, *Setting the Record Straight*, 13.

42 A.C. Edwards, *A History of Essex* (London, 1958), 10.

43 Edwards, *History of Essex*, 10.

44 G.N. Garmonsway (trans)., *The Anglo-Saxon Chronicle* (London, 1972,1982), 61.

45 Garmonsway (trans)., *The Anglo-Saxon Chronicle*, 63.

46 Garmonsway (trans)., *The Anglo-Saxon Chronicle*, 97.

47 The key work that makes this argument is Peter Sawyer, *The Age of the Vikings* (London, 1962).

48 J. Button, Walton in 1297 in the time of Alice Horlefrensche (Lulu Online Publishing, 2009), 7.

49 Jarvis, *Essex: A County History*, 50.

50 http://www.battleofmaldon.org.uk/ (accessed on 02/02/2012).

51 Jarvis, *Essex: A County History*, 51.

52 Garmonsway (trans), *Anglo Saxon Chronicle*, 145.

[53] Cnut the Great was succeeded by Harold Harefoot (1037-1040) and then Harthacnut (1040-1042) before Edward the Confessor took the throne on 8th June 1042.

[54] www.british-history.ac.uk/report.aspx?compid=35353 'Secular canons: Cathedral of St. Paul', *A History of the County of London: Volume 1: London within the Bars, Westminster and Southwark* (1909), 409-433. (accessed on 14/07/2010).

[55] Button, *Walton in 1297*, Ch.8.

[56] Jarvis, *Essex: A County History*, 52.

[57] S.H. Rigby, 'Social Structure and economic change in late medieval England' in R. Horrox and W.M. Ormrod, *A Social History of England 1200-1500* (Cambridge, 2006), 1.

[58] Rigby, 'Social Structure and economic change in late medieval England', 8-9.

[59] Button, Walton in 1297, 4.

[60] L.C. Modlen, *The Story of Walton-le-Soken* (Walton-on-the-Naze, 1955), 2.

[61] R. Pratt, *Aelduluesnasa: Kirby in History*, 6.

[62] Pratt, *Aelduluesnasa: Kirby in History*, 6.

[63] Button, Walton in 1297, 4.

[64] Jarvis, Essex: A County History, 55.

[65] Button, *Walton in 1297*, 8.

[66] P. Edwards, *Farming: Sources for Local Historians* (London, 1991), 20.

[67] Edwards, *Farming: Sources for Local Historians*, 20.

[68] H. Yelloly-Watson, *The Tendring Hundred in the Olden Time* (Colchester, 1877), 58.
In 1217 Birch Hall was held by Ermengard Makarel as a knight's fee of the Abbott of St Osyth (mentioned in Book of fees, i. 237, 241) so a transfer of ownership must have taken place at some point between 1086 and 1217. – See Button, *Walton in 1297*, footnotes end of Ch.10

[69] Palmer, *Setting the Record Straight*, 15.

[70] Palmer, *Setting the Record Straight*, 15.

[71] Garmonsway (trans.), *The Anglo-Saxon Chronicle*, 213.

[72] Garmonsway (trans.), *The Anglo-Saxon Chronicle*, 216.
Canute (who is also the patron saint of Denmark) was King of Denmark from 1080 to his death in 1086 and was one of the

many illegitimate children of King Sweyn Estridsson (see below).

[73] Sweyn of Denmark (Old Norse: Sveinn Ástríðarson) was the nephew of Cnut the Great, King of England, Denmark and Norway until 1035. Sweyn reigned as King of Denmark from 1047 to his death in 1074 and attacked England with Edgar Atheling, capturing York in 1069. However, William paid him off. Sweyn tried again in 1074 but failed.

[74] Garmonsway (trans.), *The Anglo-Saxon Chronicle*, 217.

[75] Button, *Walton in 1297*, 7.

[76] G.H. Martin and A. Williams (eds.), *Domesday Book: A Complete Translation* (London, 2003), 978.

[77] Button, *Walton in 1297*, 16.

[78] Button, *Walton in 1297*, 4.

[79] W.H. Hale, *The Domesday of St Paul's of the Year 1222* (London, 1858).

[80] Button, *Walton in 1297*, Ch. 11, 59-61.

[81] Button, *Walton in 1297*, Ch.12.

[82] VCH Essex, online text-in-progress, *Walton-le-Soken, Landownership* (2007), 1.

[83] W.H. Hale (ed), *The Domesday of St. Paul's of the year M.CC.XXII*, (London, 1858), 48.

[84] D/DAc 374. Essex Records Office.

[85] D/DAc 374. Essex Records Office.

[86] Button, *Walton in 1297*, 6.

[87] Button, *Walton in 1297*, 12.

[88] I. Johnson, *Turning Point: The story of Kirby-le-Soken, Essex* (London, 1982), 23-24.

[89] Johnson, *Turning Point: The story of Kirby-le-Soken, Essex*, 23-24.

[90] Edwards, *Farming: Sources for Local Historians*, 21.

[91] Button, *Walton in 1297*, 38.

[92] Button, *Walton in 1297*, 74.

[93] Button, *Walton in 1297*, 8.

[94] Button, *Walton in 1297*, 8.

[95] Button, *Walton in 1297*, 8. Sheeps cheese was a particular speciality for this part of Essex.

[96] Button, *Walton in 1297*, 8.

[97] H. Grieve, *The Great Tide* (Chelmsford, 1959), 5.

⁹⁸ Button, *Walton in 1297,* 19. The trees were coppiced collectively and the wood was fully replenished every 5ᵗʰ year. Thorpe, where there was more woodland, had a 7 year cycle.

⁹⁹ Button, Walton in 1297, 18.

¹⁰⁰ Button, Walton in 1297, 20 This is based on the 1297 list of labour services required by the villeins of Kirby and Horlock – '...and they are to carry from the Lord's Curia right to the ship (usque ad navem) six quarters of produce per year.'

¹⁰¹ Palmer, *Setting the Record Straight,* 25.

¹⁰²
http://www.historylearningsite.co.uk/dissolution_monasteries.htm (accessed on 25/07/10).

¹⁰³
http://www.historylearningsite.co.uk/dissolution_monasteries.htm (accessed on 07/01/12).

¹⁰⁴
http://www.historylearningsite.co.uk/dissolution_monasteries.htm (accessed on 07/01/12).

¹⁰⁵ Edward VI granted Sir Thomas Darcy the manors of Great and Little Clackton, Chiche St Osyth, Thorpe, Kirby and Walton as well as a further 6,800 acres of land. This was declared on the 4ᵗʰ April 1551.

¹⁰⁶ Senter, *A Study of the Lords of the Manor of Walton-le-Soken,* 6.

¹⁰⁷ http://www.oxforddnb.com/view/article/70579?docPos=4 (accessed on 10/01/2012) Oxford Dictionary of National Biography. Article by David Loades.

¹⁰⁸Senter, *A Study of the Lords of the Manor of Walton-le-Soken,* 7.

¹⁰⁹ Edwards, *Farming: Sources for Local Historians,* 36.

¹¹⁰ Edwards, *Farming: Sources for Local Historians,* 36.

¹¹¹
http://www.stosyth.gov.uk/default.asp?calltype=jun03elizabeth one (accessed on 14/07/2010).

¹¹² Senter, *A Study of the Lords of the Manor of Walton-le-Soken,* 6.

¹¹³ Senter, *A Study of the Lords of the Manor of Walton-le-Soken,* 6.

[114] http://www.oxforddnb.com/view/article/69349 (accessed on 09/01/2012) Oxford Dictionary of National Biography. Article by John Walter.

[115] http://www.oxforddnb.com/view/article/69349 (accessed on 09/01/2012) Oxford Dictionary of National Biography. Article by John Walter.

[116] http://www.oxforddnb.com/view/article/69349 (accessed on 09/01/2012) Oxford Dictionary of National Biography. Article by John Walter.

[117] The diary of Ralph Josselin, 1616–1683 in A. MacFarlane (ed.), British Academy, Records of Social and Economic History, ser. 3 (1976).

[118] http://www.oxforddnb.com/view/article/69349 (accessed on 09/01/2012) Oxford Dictionary of National Biography. Article by John Walter.

[119] Much information also sourced from http://www.britishhistory.ac.uk/report.aspx?compid=74279&strquery=bessie+Savage (accessed on 15/07/2010)

[120] VCH Essex, online text-in-progress, *Landownership: Kirby-le-Soken* (2007), 3.

[121] VCH Essex, online text-in-progress, *Landownership: Kirby-le-Soken* (2007), 3.

[122] Much biographical information on the Savage dynasty sourced from http://www.runcornhistsoc.org.uk/rockandclifton.html (accessed on 14/07/2010).

[123] Senter, *A Study of the Lords of the Manor of Walton-le-Soken*, 16.

[124] Modlen, *Story of Walton-le-Soken*, 7.

[125] A. Butler, *Sheep* (Winchester, 2006), 79.

[126] Senter, *A Study of the Lords of the Manor of Walton-le-Soken*, 7.

[127] Essex Recusant Society, 'Darcy and Savage' in *the Essex Recusant*, vol.27 (1985), 16.

[128] http://www.oxforddnb.com/view/article/24723?docPos=1 (accessed on 6/01/2012) Oxford Dictionary of National Biography. Article by John B. Hattendorf.

[129] P. B. Boyden, *History of Walton: Part 2, 1553-1800, The Birth of a Resort* (Walton-on-the-Naze, 1972), 37.

[130] Boyden, *History of Walton: Part 2*, 37.
[131] Boyden, *History of Walton: Part 2*, 37.
[132] Senter, A Study of the Lords of the Manor of Walton-le-Soken, 15.
[133] Senter, *A Study of the Lords of the Manor of Walton-le-Soken*, 15.
[134] http://www.oxforddnb.com/view/article/30312?docPos=2 (accessed on 10/01/2012) Oxford Dictionary of National Biography. Article by Geoffrey W. Rice.
[135] http://www.oxforddnb.com/view/article/30312?docPos=2 (accessed on 10/01/2012) Oxford Dictionary of National Biography. Article by Geoffrey W. Rice.
[136] http://www.oxforddnb.com/view/article/23647?docPos=1 (accessed on 10/01/2012) Oxford Dictionary of National Biography. Article by Roland Thorne.
[137] E.A Wood, *A History of Thorpe-le-Soken to the year 1890* (Thorpe-le-Soken, 1975), 121.
[138] 1778 general survey of 'the Mistley estate' located at the Essex Record Office – D/DFI E1.
[139] VCH Essex, online text-in-progress, *Walton-le-Soken, Landownership* (2007), 5 and 1778 general survey of 'the Mistley estate' located at the Essex Record Office – D/DFI E1.
[140] http://www.oxforddnb.com/view/article/23647?docPos=1 (accessed on 10/01/2012) Oxford Dictionary of National Biography. Article by Roland Thorne.
[141] Senter, *A Study of the Lords of the Manor of Walton-le-Soken*, 9.
[142] Mistley itself was auctioned off in lots in a series of auctions in 1844. The first auction began at noon on 9th August 1844. Mistley Hall mansion was demolished that same year and is now the site of 'Acorn Village'. Further sales took place in September 1844 and in 1845 and included sales of the docks and shipyards, warehouses, farms and land, businesses, hotels, inns, Vicarages and Rectories.
[143] The last will and testament of the Rt. Hon. Richard Rigby is located at the Essex Record Office (D/DU 5/84). In the document he grants his daughter Sarah Lucas the sum of £5000, her mother £1000, Miss Jenny Richard an annuity of £100 a year and £2000 to each of his executors for their

trouble. He granted the sum total of his estates to his heirs Martha Hale, Anne Rigby and Francis Hale-Rigby.

144 Senter, *A Study of the Lords of the Manor of Walton-le-Soken*, 13.

145 Wood, *A History of Thorpe-le-Soken*, 4

146 Senter, *A Study of the Lords of the Manor of Walton-le-Soken,* 5.

147 Senter, *A Study of the Lords of the Manor of Walton-le-Soken,* 5. Anyone interested in serious scholarship on this subject should consult E. A. Wood's book *The End of the Soken Court* (St Leonards-on-Sea, 1954).

148 Senter, *A Study of the Lords of the Manor of Walton-le-Soken,* 5.

149 VCH Essex, online text-in-progress, *Walton-le-Soken, Landownership* (2007), 6.

150 George, W and S. Vincent, 'Report of Field Meeting to Walton-on-the-Naze and Wrabness, Essex, 2.X, 1976 with Notes on the London Clay of Walton.' (2nd June 1977), p.87 – located at Frinton and Walton Heritage Trust archive

151Jarvis, *Essex: A County History*, 127.

152 A fuller account of copperas production can be found in the Victoria County History VCH Essex ii, 411-13.

153 Ford, *Land of Milk and Hunnye*, 13.

154 Ford, *Land of Milk and Hunnye*, 13.

155 W.G. Benham, *Essex Sokens: Stories of the Past* (Colchester, 1928), 20.

156 Ford, *Land of Milk and Hunnye*, 15.

157 Senter, A Study of the Lords of the Manor of Walton-le-Soken, 16.

158 Senter, *A Study of the Lords of the Manor of Walton-le-Soken,* 16.

159 VCH Essex, online text-in-progress, *Walton-le-Soken, Landownership* (2007), 6.

160 ERO D/DBm M578.

161 Senter, *A Study of the Lords of the Manor of Walton-le-Soken,* 18.

162 VCH Essex, online text-in-progress, *Walton-le-Soken, Landownership* (2007), 6.

163 ERO D/DHw E9.

[164] ERO D/DHw E9.

[165] A. Young, *General View of the Agriculture of the County of Essex Vol II* (London, 1807), 417.

[166] Deeds of Walton Hall, held by the solicitors of the Eagle family.

[167] ERO D/APsW 1/31

[168] ERO D/APsW 1/31

[169] ERO D/APsW 1/31 *and* Diaries of Richard Stone 1810-1832, courtesy of the owners of the diaries

[170] Diaries of Richard Stone 1810-1832, courtesy of the owners of the diaries

[171] ERO D/APsW 1/31

[172] Diaries of Richard Stone

[173] Diaries of Richard Stone

[174] Diaries of Richard Stone

[175] Diaries of Richard Stone

[176] Diaries of Richard Stone

[177] Diaries of Richard Stone. *A 'horse' appears to have been a type of scaffolding used for supporting stonework in water during construction work on sea walls.*

[178] Diaries of Richard Stone

[179] Diaries of Richard Stone

[180] Diaries of Richard Stone 1810-1832

[181] Diaries of Richard Stone

[182] Deeds of Walton Hall, held by the solicitors of the Eagle family.

[183] Deeds of Walton Hall.

[184] Deeds of Walton Hall.

[185] ERO D/DU 493/12.

[186] ERO D/DBm M578.

[187] 1986 Naze protection society booklet, 2.

[188] Ford, *Land of Milk and Hunnye*, 92.

[189] Ford, *Land of Milk and Hunnye*, 95.

[190] J.W. & F.D. Eagle farm deed documents.

[191] J.W. & F.D. Eagle farm deed records. Hedge End Island was used for sheep grazing but was soon disposed of, being sold to Messrs Wilson and Nicholson on 9th August 1921.

[192] Ben Eagle, Interview, Daphne Eagle (Kirby-le-Soken, 2/09/2014).

[193] 1933 Agreement between Trinity House and Naze Golf Club (located at Frinton and Walton Heritage Trust archive).

[194] 1986 Naze protection society booklet, 3.

[195] 1986 Naze protection society booklet, 3.

[196] Letter to Messrs Cross, Guest, Lowden and Hasell from L.B. Grimshaw dated 22nd February 1962.

[197] Minutes of a Meeting of the Walton Links Estate Sub-committee (28th May 1965).

[198] 1986 Naze protection society booklet, 3. The Council were then known as the Frinton and Walton Urban District Council. Tendring District Council now own the area of land as the successor organisation.

[199] Ben Eagle, Interview, Mike Todd, (Walton-on-the-Naze, 27.09.2014).

[200] Letter of objection from Mr J K Weston to proposals to increase number of berths in Hamford Water in the 1980s.

[201] P. Ford, *Tendring peninsula: Land of Milk and Hunnye* (Bristol, 1988), 3.

[202] B. Oxborrow. 'Our President Remembers' (May 1995) – Frinton and Walton Heritage Trust archive

[203] Button, *Walton in 1297*, 21.

[204] Button, *Walton in 1297*, 21.

[205] Button, *Walton in 1297*, 21.

[206] Button, *Walton in 1297*, 21.

[207] Button, *Walton in 1297*, 21.

[208] Button, *Walton in 1297*, 22.

[209] Button, *Walton in 1297*, 22.

[210] Button, *Walton in 1297*, 22.

[211] Button, *Walton in 1297*, 22.

[212] Button, *Walton in 1297*, 22.

[213] VCH Essex, online text-in-progress, *Walton-le-Soken, Landownership* (2007), 4.

[214] Button, *Walton in 1297*, 68.

[215] VCH Essex, online text-in-progress, *Walton-le-Soken, Landownership* (2007), 4.

[216] VCH Essex, online text-in-progress, *Walton-le-Soken, Landownership* (2007), 4.

[217] Boyden, *History of Walton: Part 2*, 27. J. Norden, 1594, *Speculi Britanniae Pas: The Description of Essex* (London, Camden Society, 1840):

[218] Mid 19th Century advertising booklet entitled 'A Descriptive Account of Walton-on-the-Naze' printed by Edward Benham of Colchester, 17

[219] Button, *Walton in 1297,* 66.

[220] Button, *Walton in 1297,* 66.

[221] Button, *Walton in 1297,* 70.

[222] Button, Walton in 1297, 65.

[223] Button, Walton in 1297, 65.

[224] Button, *Walton in 1297,* 65.

[225] Button, *Walton in 1297,* 65.

[226] U. Bloom, *The House that Died Alone* (London, 1964), 11.

[227] Sale Notice of Walton Hall (1801). Available at Frinton and Walton Heritage Trust archive, Walton Maritime Museum.

[228] ERO DHw E9

[229] ERO D/DU 954

[230] ERO D/DHw E7.

[231] 1820 valuation located at the Essex record office – ERO D/DHw E7.

[232] http://www.bbc.co.uk/news/uk-england-essex-23879430 (accessed 12.01.14)

[233] Memoirs of J.W Eagle (1922-2005)

[234] Memoirs of A.H.B Franklin regarding Walton Hall, dated January 2005.

[235] Memoirs of A.H.B Franklin

[236] Memoirs of A.H.B Franklin

[237] J.W & F.D Eagle, Walton Hall farm deeds.

[238] Palmer, Setting the Record Straight, 37.

[239] Goldmer Gat was first surveyed in 1628 as expressed in a letter from Trinity House to the Privy Council 24th Dec 1628. Located at http://www.british-history.ac.uk/report.aspx?compid=63925&strquery=Goldmer+g at

[240] ERO D/DU 954

[241] P. Morant, *The History and Antiquities of the County of Essex* (London, 1768), 91.

[242] D. Defoe, *A Tour Through the Whole Island of Great Britain: Divided Into Circuits or Journies* (D. Browne et al, 1762), 18.

[243] Palmer, *Setting the Record Straight*, 37.

[244] Defoe, *A Tour Through the Whole Island of Great Britain*, 15.

[245] R. O'Neil, The Naze Tower at Walton: The First Line of Defence (Bognor Regis, 2006), 6.

[246] The Bedfordites were an 18th Century political faction, led by John Russell, 4th Duke of Bedford. Notable members included John Montagu, 4th Earl of Sandwich; Granville Leveson-Gower, 2nd Earl Gower; Rigby, Thomas Thynne, 3rd Viscount Weymouth; Edward Thurlow; and George Spencer, 4th Duke of Marlborough.

[247] Palmer, *Setting the Record Straight*, 92.

[248] A Martello Tower is a circular masonry defensive fort of up to twelve metres (forty feet) high, holding garrison for fifteen to twenty five men. They were built during the time of the Napoleonic Wars to guard against the French invasion force.

[249] T. Wilmshurst, *A Descriptive Account of Walton-on-the-Naze* (London, 1860), 11.

[250] Frinton and Walton Heritage Trust archive, Walton

[251] The book is entitled *The Naze Tower at Walton: The first line of defence* and is available through Woodfield Publishing, Bognor Regis.

[252] G. Rayner, *Seaside Front Line* (Medlesham, 1996), 28.

[253] Rayner, *Seaside Front Line*, 29.

[254] Rayner, *Seaside Front Line*, 55 and 113.

[255] Naze Tower pamphlet entitled *'Naze Tower: Past, Present and Future'* (2010).

[256] Naze Tower pamphlet entitled *'Naze Tower: Past, Present and Future'* (2010).

[257] The reestablishment of a 'tea rooms' has precedent from the 'tea rooms' supposedly established by the Rt. Hon. Richard Rigby in the eighteenth century. However, tea held today is a more civilised air than in the age of Rigby.

[258] H. Grieve, *The Great Tide* (Chelmsford 1959), 1.

[259] Morant, The History and Antiquities of the County of Essex, 91.

[260] 1960s pamphlet on coastal erosion at the Naze.
[261] Gowen, P., Why Canute Failed: A Treatise on Sea Defences located at http://www.marinet.org.uk/coastaldefences/canute.html (written Nov 2005, updated April 2008, sourced 1/10/11).
[262] Dalton, W.H., 'Walton and Frinton in 1902' in *Essex Naturalist* vol.12 (1902), 217-221.
[263] IPCC, Climate Change 2014: Impacts, Adaptation and Vulnerability (Fifth Assessment Report, 2014), 368-372 available at http://www.ipcc.ch/pdf/assessment-report/ar5/wg2/WGIIAR5-Chap5_FINAL.pdf
[264] The marshes were originally drained in the late seventeenth century by a team of Dutch engineers under the leadership of Nicholas van Cropenbourgh. Following 1953, further amounts of spoil were transferred from the marshes to the walls themselves.
[265] H. Grieve, The Great Tide (Chelmsford, 1959), 7.
[266] J.M. Stratton, Agricultural Records A.D 220-1968 (London, 1969), 15-18.
[267] Button, Walton in 1297, 70.
[268] Button, Walton in 1297, 71.
[269] Grieve, The Great Tide, 6.
[270] Grieve, The Great Tide, 8.
[271] Button, Walton in 1297, 71.
[272] Lamb, H.H., *Climate, History and the Modern World* (London, 1982), 25.
[273] Grieve, The Great Tide, 14.
[274] Grieve, The Great Tide, 21.
[275] Grieve, The Great Tide, 23.
[276] Grieve, The Great Tide, 31.
[277] Button, Walton in 1297, 72.
[278] Button, Walton in 1297, 72.
[279] http://www.historyhouse.co.uk/placeW/essexw05a.html (accessed on 15/07/2010).
[280] Grieve, The Great Tide, 40.
[281] Diaries of Richard Stone
[282] Grieve, The Great Tide, 42.
[283] Grieve, The Great Tide, 42.
[284] Grieve, The Great Tide, 48.

285 Grieve, The Great Tide, 48.

286 Grieve, The Great Tide, 58.

287 Grieve, The Great Tide, 92.

288 Grieve, The Great Tide, 127.

289 Grieve, The Great Tide, 177, 807.

290 Grieve, The Great Tide, 814.
Hilda Grieve's great study in to the 1953 disaster is highly recommended. Reference to this specific topic: H. Grieve, The Great Tide (Chelmsford, Essex County Council, 1959).

291 Entry by J.W Eagle from 1953 farm diary.

292 Save the Naze was established as a registered charity in 1975 with the Naze Protection Society holding its first official meeting on 21st January 1981.

293 Naze Protection Society Booklet entitled 'The Naze at Walton' (Walton-on-the-Naze, 1987), 4.

294 Naze Protection Society booklet 1981

295 Naze Protection Society Booklet entitled 'The Naze at Walton' (Walton-on-the-Naze, 1987), 5.

296 Naze Protection Society Booklet entitled 'The Naze at Walton' (Walton-on-the-Naze, 1987), 5.

297 Naze Protection Society Booklet entitled 'The Naze at Walton' (Walton-on-the-Naze, 1987), 8.

298 'The Naze - Notes on meeting held on 25th October 1996' – Frinton and Walton Heritage Trust archive

299 Naze Protection Society, 'Naze News', issue 14 (Summer 2005), 4.

300 Naze Protection Society, 'Naze News', issue 13 (Spring 2005), 1.

301 http://www.nazeprotectionsociety.org.uk/cragwalk.htm (accessed on 07.01.12).

302 http://www.nazeprotectionsociety.org.uk/cragwalk.htm (accessed on 07.01.12).

303 D. Paine (ed), The History and Work of the Walton and Frinton Lifeboat (Walton, 1975), 7.

304 J. Steer, Walton and Frinton Lifeboat: A Station History 1884-2005 (Lavenham, 2005), 3.

305 Steer, Walton and Frinton Lifeboat, 6.

306 Steer, Walton and Frinton Lifeboat, 9.

307 Steer, Walton and Frinton Lifeboat, 11.

[308] Paine (ed), The History and Work of the Walton and Frinton Lifeboat, 21.

[309] Paine (ed), The History and Work of the Walton and Frinton Lifeboat, 51.

[310] The other naval forts were at Sunk Head, Tongue Sands and Knock John.

[311] Email correspondence 2012.

[312] Button, Walton in 1297, 71.

[313] Naze Protection Society , 'The Endangered Naze: A Guide to Flora, Fauna and Bird Life to be found there' (Walton, 1981)

[314] Letter to Messrs Gross, Guest, Lowden and Hasell from L.B. Grimshaw dated 22nd February 1962.

[315] R.J. Nicholls, and S. Rupp, *Managed Realignment of Coastal Flood Defences: A Comparison between England and Germany* (London, University of Middlesex, 2002) available at http://www.survas.mdx.ac.uk/public11.htm

Lightning Source UK Ltd.
Milton Keynes UK
UKOW04f1327020916

281984UK00001B/108/P